Disclaimer

The publisher of this book is by no way associated with the National Institute of Standards and Technology (NIST). The NIST did not publish this book. It was published by 50 page publications under the public domain license.

50 Page Publications.

Book Title: Limits to the Effectiveness of Metal-Containing Fire Suppressants. Final Technical Report (NISTIR 7177)

Book Author: Gregory T. Linteris

Book Abstract: This report reviews the literature on metal inhibition of flames and identifies metal species with potential as fire suppressant additives. To provide a basis for discussion, the detailed mechanism of inhibition of iron is reviewed, and the reasons for its loss of effectiveness are described. The demonstrated flame inhibiting properties of other metals is then discussed, followed by a description of the potential loss of effectiveness for these other metals. The production ban on the widely used and effective halon fire suppressants due to their ozone depletion potential, has motivated an extensive search for replacements. Metal containing compounds have attracted attention- especially for unoccupied spaces-because of their extraordinary effectiveness in some configurations. For example, $Fe(CO)_5$ has been found to be up to eighty times more effective than CF_3Br at reducing the overall reaction rate in premixed methane-air flames, when added at low concentration. Unfortunately, it has also been found to produce condensed-phase particles which reduce its effectiveness for co-flow diffusion flames. Hence, it is of interest to identify other metal compounds which may be strong flame inhibitors and then to assess their potential for loss of effectiveness through condensation. To achieve this goal, the present report provides background on current understanding of metal inhibition of flames, identifying metals with fire suppression potential. The inhibition mechanism of the iron is described, and the followed by a description of the reasons why it losses its effectiveness in some flame systems. The equivalent flame inhibiting species of other metal agents is then discussed, and evidence for any potential loss of effectiveness for these other metals is assembled and discussed.

Citation: NIST Interagency/Internal Report (NISTIR) - 7177

Keyword: metals; fire suppression; additives; iron; effectiveness; halons; ozone; halon alternatives; reaction rate; premixed flames; diffusion flames; flame extinguishment; condensation; flame retardants; ignition; combustion; nozzles

NISTIR 7177

Limits to the Effectiveness of Metal-Containing Fire Suppressants

Gregory T. Linteris

National Institute of Standards and Technology
Technology Administration, U.S. Department of Commerce

NISTIR 7177

Limits to the Effectiveness of Metal-Containing Fire Suppressants

Gregory T. Linteris
Fire Research Division
Building and Fire Research Laboratory

October 2004

U.S. DEPARTMENT OF COMMERCE
Donald L. Evans, Secretary
TECHNOLOGY ADMINISTRATION
Phillip J. Bond, Under Secretary of Commerce for Technology
NATIONAL INSTITUTE OF STANDARDS AND TECHNOLOGY
Arden L. Bement, Jr., Director

Final Technical Report

Limits to the Effectiveness of Metal-Containing Fire Suppressants

Date: 10/12/2004

Gregory Linteris

Building and Fire Research Laboratory
Fire Research Division
100 Bureau Dr. Stop 8652
Gaithersburg MD 20899-8652

October 2004

Final Technical Report, March 1, 2004 – July 31, 2004

The views and conclusions contained in this document are those of the authors and should not be interpreted as representing the official policies, either expressed or implied, of the Strategic Environmental Research and Development Program, National Institute of Standards and Technology, or any other part of the U.S. Government.

Sponsored by:
The Department of Defense
Strategic Environmental Research and Development Program

Abstract

This report reviews the literature on metal inhibition of flames and identifies metal species with potential as fire suppressant additives. To provide a basis for discussion, the detailed mechanism of inhibition of iron is reviewed, and the reasons for its loss of effectiveness are described. The demonstrated flame inhibiting properties of other metals is then discussed, followed by a description of the potential loss of effectiveness for these other metals.

Introduction

The production ban on the widely used and effective halon fire suppressants due to their ozone depletion potential, has motivated an extensive search for replacements (Gann, 2004). Metal containing compounds have attracted attention—especially for unoccupied spaces—because of their extraordinary effectiveness in some configurations. For example, $Fe(CO)_5$ has been found to be up to eighty times more effective than CF_3Br at reducing the overall reaction rate in premixed methane-air flames, when added at low concentration (Reinelt and Linteris, 1996). Unfortunately, it has also been found to produce condensed-phase particles which reduce its effectiveness for co-flow diffusion flames (Linteris and Chelliah, 2001). Hence, it is of interest to identify other metal compounds which may be strong flame inhibitors and then to assess their potential for loss of effectiveness through condensation. To achieve this goal, the present report provides background on current understanding of metal inhibition of flames, identifying metals with fire suppression potential. The inhibition mechanism of the iron is described, and the followed by a description of the reasons why it losses its effectiveness in some flame systems. The equivalent flame inhibiting species of other metal agents is then discussed, and evidence for any potential loss of effectiveness for these other metals is assembled and discussed.

Background

There has been much work, in a number of different applications, on the effect of metal compounds on the high-temperature oxidation reactions of hydrocarbons. The interest in the present work is on their potential as fire suppressants, as well as evidence for and mechanisms of a loss of effectiveness that they might demonstrate for certain fire suppression applications. Since their inhibition, as well as any loss of effectiveness, may be governed by consistent mechanisms between the different applications, we explore all relevant chemical systems that might have bearing on their performance as fire suppressants.

The behavior of metal compounds in flames has been studied with regard to several flame phenomena. In the most obviously relevant work, metal compounds have been added to flames in a number of screening tests, aimed specifically at assessing their potential as fire suppressants. In flames, screening tests have has been performed to understand their influence on flame speed (premixed flames) and extinction (diffusion flames), as well as their effect on ignition. Flame studies have also been used to understand the detailed mechanism of inhibition of metal agents, either by providing direct experimental data on species present in the flame zone, or by validating numerical models which are then used to calculate the flame structure; either of these approaches are then used to develop an understanding of the relevant chemical kinetic mechanisms. Studies of engine knock suppression by metals provided much early data on metals in flames. Since engine knock is known to occur from the rapid pressure rise (and subsequent detonations) inside an engine cylinder caused by too fast reaction of the homogeneous charge of fuel and air, the mechanisms of engine knock reduction may have clear relevance to fire suppression—in which the goal is again to reduce the overall reaction rate with the addition of the suppressing agent. In engines—as well as in heating applications—research has also been directed at understanding their efficacy at soot reduction. Although both soot formation and the overall reaction rate of flames are known to be related to the location and concentration of radicals in the flame, the effects of metals on soot formation are not reviewed here.

Other systems have been used to understand metal chemistry in flames. A large amount of fundamental work has been done with premixed atmospheric-pressure flat flames. In this system, the flat flame provides a nearly one-dimensional system, and the region above the flame (i.e., downstream of the main reaction zone) provides a long residence time, high temperature region for radical recombination. The H-atom concentration typically is measured with the Li-LiOH method (described below), and the additive's effect on the radical recombination rate is determined. Ignition studies using flames for screening tests have been reported. In addition, much detailed fundamental understanding has come from shock-tube studies and flash-photolysis studies in reaction vessels. Some studies of fire retardants are also relevant to fire suppression mechanisms of metals. For example, when the fire retardant works by suppressing the gas-phase reactions (and the subsequent heat release and heat feedback to the solid sample), the mechanisms are directly relevant to fire suppression. Finally, after-burning in rocket nozzles provided motivation to understand metal-catalyzed radical recombination reactions, and modeling studies have been performed for those systems.

From these studies, it is clear that metals can have a profound effect on flame chemistry. Further, their effectiveness in these varied applications may well be related. In any event, data from each of the applications can provide insight into possible metals for application to fire suppression as well as provide fundamental data useful for predicting their performance in a range of applications. Work investigating the effect of metals for each of the applications is described below.

Engine Knock

Agents that reduce engine knock may also be effective flame inhibitors, and it is useful to examine the literature of engine knock to search for possible moieties. Engine knock is the onset of detonation waves in an engine cylinder brought about by the homogeneous ignition of the end-gas region of highly compressed and heated fuel and air. The effect of some agents in reducing knock has been known since the 1920s, including compounds of bromine, iodine, tellurium, tin, selenium, iron, and lead, as well as aniline (Kovarik, 1994). Tetraethyl lead (TEL) very early became the anti-knock agent of choice. While much subsequent research was performed to understand its mechanism of knock reduction, the exact mechanism for this agent remains an unsolved problem in combustion research (perhaps because leaded fuels were later banned due to their poisoning effect on exhaust catalytic converters). Although much progress was made, the researchers divided into two camps: those endorsing a heterogeneous mechanism

(Walsh, 1954) and those promoting a homogeneous radical recombination mechanism (Erhard and Norrish, 1956).

Several known effects of lead in engines support the heterogeneous mechanism. Muraour (1925) appears to have been the first to propose chain-breaking reactions on the surface of a colloidal fog formed from TEL. The particle cloud was subsequently shown to be composed of PbO, which is the active species (Chamberlain and Walsh, 1952). Since a strong influence of PbO coatings on reaction vessel walls has also been observed (Chamberlain and Walsh, 1952), a heterogeneous mechanism of PbO was assumed. Although other results support the heterogeneous mechanism, the evidence is somewhat circumstantial. The known *metallic* antiknock compounds (tetraethyl lead, tellurium diethyl, iron pentacarbonyl, nickel tetracarbonyl, (Walsh, 1954)) all produce a fog of solid particles. The alkyls of bismuth, lead, and thallium are anti-knocks, but those of mercury (which does not form particles) are not (Cheaney et al., 1959). Richardson et al. (1962) showed that carboxylic acids increase the research octane number of TEL in engines, and argued that they reduced agglomeration of the PbO particles in the engine end-gas, but acknowledged that their arguments were qualitative. Zimpel and Graiff (1967) sampled end gases in a fired engine for transmission electron microscopy (TEM). They claimed that 1000 nm diameter particles formed prior to the arrival of the flame, proving that the effect of TEL was heterogeneous. Commenters to the paper pointed out, however, that the particles could be forming in the sampling system, that the effect of lead extenders was not captured in the particle morphology, and that even if particles form, the gas-phase species can also be present, and it is thus not precluded that they are doing the inhibition. Based on the work of Rumminger and Linteris (2000c), even if particles form, they can re-evaporate in the hot region of the flame if they are small enough. Finally, as described by Walsh (1954), a major shortcoming in the work with TEL is that it is easily absorbed on the way to the reaction vessel, so that it is difficult to know how much TEL actually makes it to the flame. As subsequently described by Linteris and co-workers, the efficiency (Reinelt and Linteris, 1996; Rumminger et al., 1999; Rumminger and Linteris, 2000a; Linteris et al., 2000a; Linteris et al., 2002) and particle formation (Rumminger and Linteris, 2000c; Rumminger and Linteris, 2002) of organometallic agents are strongly affected by the volume fraction of the metal compound.

In later work, Kuppo Rao and Prasad (1972) claimed to prove the heterogeneous mechanism of lead anti-knock agents. They inserted copper fins coated with PbO into the cylinder of an engine, or injected 30 µm particles into the air stream, and found antiknock effects. They interpreted these results as evidence that the mechanism is heterogeneous. Nonetheless, they did not measure for the presence of gas-phase lead compounds, so a homogeneous mechanism cannot be ruled out. For the eleven lead compounds tested, they found the effectiveness to vary by a factor of about six, and found similarly sized particles of CuO_2, CuO, $CuCl_2$, $NiCl_2$, and $SnCl_2$ to have equal effectiveness which was less than any of the lead compounds.

The early and strong evidence for a homogeneous gas-phase inhibition mechanism of TEL was developed by Norrish (1956). Using flash photolysis of mixtures of acetylene, amyl nitrite, and oxygen in a reaction vessel, with and without TEL, the absorption and emission spectra of amyl nitrate, OH, Pb, PbO, NO, CN, CH, and TEL were obtained as a function of reaction progress. The reactants were chosen since, in the absence of anti-knock compounds, they showed the strong homogeneous detonation characteristic of engine knock. The researchers found that the induction time increased linearly with TEL addition at low TEL partial pressures, but that the effectiveness dropped off at higher pressures (a result described subsequently for the flame inhibitors $Fe(CO)_5$ (Jost et al., 1961; Reinelt and Linteris, 1996), $SnCl_4$, (Lask and Wagner, 1962; Linteris et al., 2002), and MMT (Linteris et al., 2002)). After decomposition of TEL, Pb was present in low concentration, followed by large amounts of PbO and OH , which subsequently dropped off. With TEL addition, the formation of OH was retarded, and the increase in OH emission was smoother and more well-behaved. No particles were reported. Norrish et al. described the action of TEL as a two-stage homogeneous gas-phase reaction mechanism. In the first stage, TEL reacts with the peroxide and aldehyde intermediates (characteristic of the end of the cool flame regime of alkane combustion), thus reducing the availability of these species for initiating the well-known second stage of the combustion. Gas-phase PbO from the TEL then reacts with the chain-carrying intermediates to reduce the rate of heat release, slow the temperature rise, and reduce detonation. It is noteworthy that many of the features subsequently described by Linteris and Rumminger (2002) as necessary for flame inhibition by iron compounds (effectiveness in the absence of particles, decreasing effectiveness with higher inhibitor concentration, key role of the metal monoxide species, and the necessary coexistence of OH and metal monoxide species) were shown in the 1950s by Norrish and co-workers to be necessary for effective knock reduction by TEL. Although it was of the highest quality, the findings of Norrish and co-workers were discredited by the engine community largely because the tests were not done in engines, or with typical fuels. This work predated most of the important flame inhibition work of the 1970s by 20 years.

As a prelude to the description of the flame screening tests, it is useful to describe the different terminology often used to describe the extinguishing of a flame. *Flame inhibition* usually refers to a weakening of a flame, that is, a lowering of the overall exothermic reaction rate in the flame. This weakening may or may not lead to extinguishment, depending upon the flow field in which the flame exists. In contrast, the terms *fire suppression, flame extinguishment,* or *flame extinction* are often used to refer to the case in which the flame has been weakened to the point where it can no longer stabilize in the relevant flow field. Flame quenching refers to flame extinguishment for which heat losses to a surface was a precipitating factor.

Flame Screening Tests

Papers describing the results of screening tests clearly demonstrated the superior effectiveness of some metals as flame inhibitors. The seminal work of Lask and Wagner (1962) investigated the efficiency of numerous compounds for reducing the burning velocity of premixed Bunsen-type hexane-air flames stabilized on a nozzle burner. They found the metal halides $SnCl_4$ and $TiCl_4$ to be quite effective at low volume fraction (and $GeCl_4$ about a factor of two less then these). In unpublished work, cited by others (Morrison and Scheller, 1972), Lask and Wagner provide the measured flame speed reduction with addition of $SbCl_3$ to hexane-air flames, showing it to be about twice as effective as CF_3Br. Further investigating the efficiency of metal compounds, they tested iron pentacarbonyl $Fe(CO)_5$, tetraethyl lead TEL $Pb(C_2H_5)_4$, and chromyl chloride CrO_2Cl_2 and found these to be spectacularly effective, with volume fractions of only 170 µL/L and 150 µL/L required for the first two to reduce the burning velocity the hexane-air flames by

30 % (although not quantified, they believed the effectiveness of CrO_2Cl_2 to be even higher). They categorized the compounds they tested into two classes: the halogens, and the much more effective transition metals. Miller et al (1963), using a burner which produced conical flames, tested eighty compounds for reducing the burning velocity of premixed hydrogen-air flames (fuel-air equivalence ration ϕ of 1.75). They found the most effective to be: tetramethyl lead TML $Pb(CH_3)_4$, $Fe(CO)_5$, $TiCl_4$, $SnCl_4$, $SbCl_5$, and TEL (in that order), with TEL only slightly better than CF_3Br, and TML about 11 times better. A limitation of this work, however, is that the agents were tested at only a single volume fraction. Since the efficiency of many agents is known to vary with their concentration (Reinelt and Linteris, 1996), such an approach can skew the relative performance rankings. It should also be noted that in the hotter, faster-burning hydrogen flames, the propensity to form particles from the metal oxides and hydroxides will be less than in the slower, cooler hydrocarbon-air flames of other studies (see Rumminger et al. (2000c)). An additional flame screening test was performed by Miller (Miller, 1969), who tested fifteen compounds added to low-pressure (1.01 kPa) premixed and diffusion flames, and found tin ($SnCl_4$), phosphorus ($POCl_3$), titanium ($TiCl_4$), iron ($Fe(CO)_5$), tungsten (WF_6), and chromium (CrO_2Cl_2) to have some promise.

Two screening tests involved the inhibition of propagating premixed flames through clouds of small solid particles of inhibitor. Rosser et al. (1963) added metal salts as dispersions of fine particles (2 μm to 6 μm diameter) to premixed methane-air flames. They used the reduction in upward flame propagation rate through a vertical tube as a measure of the inhibition effect. They correctly described the effects of flame speed and particle diameter on the particle heating rate, and together with the volatilization rate for each compound, assessed the fraction of the particle which was vaporized. They quantified the effect of particle size on inhibition effectiveness, and showed that while additional agent decreased the flame speed, the effectiveness eventually saturated (i.e., beyond a certain additive mass fraction, additional inhibitor had a greatly reduced effect on the flame speed). They postulated a homogeneous gas-phase inhibition mechanism involving H, O, and OH radical recombination reaction with the metal atom and its hydroxide. This mechanism had many of the features of subsequently described mechanisms (Cotton and Jenkins, 1971; Bulewicz et al., 1971; Bulewicz and Padley, 1971a; Linteris et al., 2002) which are now believed to be correct. They also correctly understood that catalytic radical recombination relies upon a super-equilibrium concentration of radicals (commonly present in premixed flames and diffusion flames at higher strain), and that this provides an upper limit to the chemical effect of catalytically acting agents. Finally, they suggested that adding an inert compound as a co-inhibitor can overcome this limitation. A number of alkali metal sulfates, carbonates, and chlorides were tested, as well as cuprous chloride CuCl, which was found to be about twice as effective as Na_2CO_3 (after correcting for the larger size of the CuCl particles.

A later study involving premixed flames with particles was performed by deWitte et al. (1964). In it, relatively large particles (100 μm diameter) were electrostatically suspended in a tube and then injected into a downward facing premixed Bunsen-type flame, and their effect on the flame temperature, burning velocity, and extinction condition was measured. Various barium, sodium, and potassium compounds were tested, as well as $AlCl_2$, $CuCl_2$, and PbO. The authors noted a thermal and chemical effect of the particles, and assumed that the chemical effect was due to recombination of chain-carrying radicals on the particle surfaces. The authors estimated that for these large particles, little of the particles could be vaporized. Nonetheless, it is not possible to separate the heterogeneous and homogeneous inhibition effects from their data. The authors found particles of $CuCl_2$ and PbO to be about two and eight times as effective as particles of Na_2CO_3, and surprisingly, found $AlCl_3$ to be about three times as effective.

A flame screening test incorporating many compounds was also performed by Vanpee and Shirodkar (1979) to test the relative effectiveness of metal salts. The metal acetonates and acetylacetonates were dissolved in ethanol, and fine droplets of the metal salt solutions were sprayed into the air stream of a counterflow diffusion flame over an ethanol pool. The inhibition effect was quantified, at a given air flow velocity (i.e., strain rate) as the change in the oxygen mole fraction at extinction caused by addition of the inhibitor, normalized by the inhibitor mole fraction (Mole merit number = $(X_{O_2,ext} - X_{O_2,ext,i})/X_i$, in which $X_{O_2,ext,i}$ and $X_{O_2,ext}$ are the oxygen mole fractions required for extinction, with and without added inhibitor, and Xi is the mole fraction of inhibitor). Their results are depicted in Figure 1, which shows the metal compounds tested, listed from most effective to least. The maximum and minimum values of the mole merit number are listed for the range of oxidizer velocities of the tests (50 cm/s to 60 cm/s). As the figure shows, the metal salts of Pb, Co, Mn, Fe, and Cr all show some inhibition effect. Interestingly, $Fe(CO)_5$ was not as effective as iron acetylacetonate, and for lead, the acetonate was about twice as effective the acetylacetonate. It should be noted, however, that the interpretation of the present data are complicated by several effects. First, changing the oxygen mole fraction changes the temperature, which can change the effectiveness of an agent, as subsequently described in refs. (Reinelt and Linteris, 1996; Rumminger et al., 1999; Rumminger and Linteris, 2000b; Rumminger and Linteris, 2000a; Macdonald et al., 2001). Since the air stream velocity is changed while holding the nebulizer flow constant, the ethanol concentration changes in these partially premixed diffusion flames. Adding changing amounts of a fuel species (i.e., the carrier ethanol) in the air stream, changes the flame location and the scalar dissipation rate for a given strain rate (i.e., air flow velocity), so that the extinction condition is modified (as discussed in (Fallon et al., 1996) and (MacDonald et al., 1998)). The size of the residual particle (which will vary from agent to agent) could affect its ability to vaporize in the flame, affecting the indicated efficiency. Finally, for metal agents that condense, their marginal effectiveness is a very strong function of the concentration at which they are added. Hence, without knowing what the additive mole fraction is, it is difficult to cross compare the effectiveness of the different agents. For these reasons, it seems appropriate to consider the results shown in Figure 1 as qualitative, rather than quantitative. For example, subsequent studies (Babushok and Tsang, 2000) have rated iron as about ten times as effective as sodium (in contrast to the results of Vanpee and Shirodkar, which show the acetylacetonate of iron to be only about 20 % more effective than that of sodium.

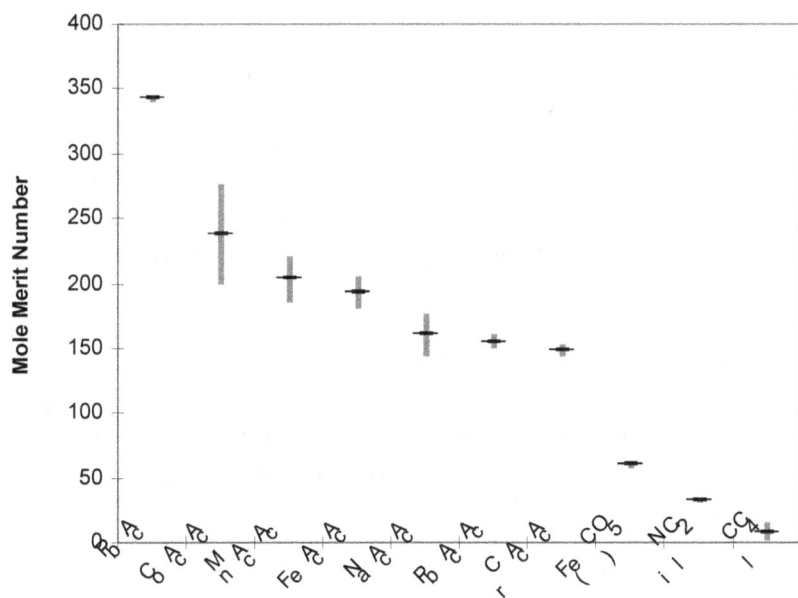

Figure 1 – Mole merit number of metal compounds for the oxidizer velocities in the range of 50-60 cm/s (from (Vanpee and Shirodkar, 1979)).

Radical Recombination above Premixed Flames

Much of the understanding of the homogeneous gas-phase flame inhibition by metals came from studies of H-atom recombination rates above premixed, fuel rich, $H_2 - O_2 - N_2$ burner-stabilized premixed flames. The techniques, pioneered by Sugden and co-workers (James and Sugden, 1956; Buelwicz et al., 1956; Bulewicz and Sugden, 1956a), involve absorption spectroscopy for OH, and the Li/LiOH technique for H atom. (In the latter technique, the strong absorption lines of Li and Na are simultaneously measured above a flat flame. With the assumptions that Na is present in the flame only in atomic form and that Li is present only as Li and LiOH, the measured ratio of the Li to Na absorption together with the known equilibrium constant for the reaction $Li + H_2O = LiOH + H$ provides [H].) In some of their early relevant work they determined the dominant metal species in H_2-O_2-N_2 premixed flames above Meker burners with added dilute sprays from aqueous salts of copper and manganese. Copper was found to exist mostly as Cu in the flame, and the dissociation constants of CuH (Bulewicz and Sugden, 1956b) and CuOH (Bulewicz and Sugden, 1956a) were determined from their concentrations above the flame at different temperatures. Similar results were subsequently obtained for MnO and MnOH (Padley and Sugden, 1959).

In a detailed study of chromium, Bulewicz and Padley (1971b) added μL/L levels of Cr (from either chromium carbonyl or aqueous sprays of chromium salts) to a premixed, fuel rich, flat-flame burner of $H_2+O_2+N_2$, and measured the catalytic radical recombination by the chromium compounds. They identified the active chromium species as Cr, CrO, CrO_2, and $HCrO_3$ and also detected solid particles, which appeared to have equivalent black-body temperatures up to 500 K higher than the gas. They estimated an <u>upper limit</u> for the rate of heterogeneous radical recombination on the particle surfaces, and estimated that, at an upper limit, it was the same order as the natural un-catalyzed recombination rate in the flame. They also found that Cr showed measurable catalytic radical recombination even when added at volume fractions of about 1μL/L—at which heterogeneous particle catalysis cannot be contributing. They cite Jenkins (1969, personal communication) as first showing the catalytic effect of metals (Ca, Sr, Ba) on radical recombination in flames. Interestingly, their data show a saturation effect in the catalytic radical recombination by Cr species, and their analyses show that it is due—not to condensation of the active gas-phase Cr-containing species to particles—but to reduction in the available radicals to recombine.

In continuing work, Bulewicz and coworkers (Bulewicz and Padley, 1971a) (Bulewicz et al., 1971) studied the catalytic effects of twenty-one metal species for recombining chain-carrying flame radicals present at super-equilibrium levels above premixed, fuel rich, flat flames of $H_2+O_2+N_2$ at 1860 K. Table 1 shows the ratio of the catalyzed to un-catalyzed recombination rate for H atom caused by each of the metals, added at 1.3 μL/L. The cut-off value for this ratio was

Table 1 - Catalytic efficiency of different metals in promoting radical recombination.
Values of k_{obs}/k_{uncat} (T = 1860°K; X[M] = 1.3 μL/L; H2/O2/N2 = 3/1/6).

Strong effect		Some effect		No effect	
Cr	2.8	Co	1.1	V	1
U	1.82	Pb	1.1	Ni	1
Ba	1.75	Zn	1.07	Ga	1
Sn	1.6	Th	1.06	Cl	1
Sr	1.35	Na	1.04		
Mn	1.3	Cu	1.04		
Mg	1.25	La	1.04		
Ca	1.25				
Fe	1.2				
Mo	1.16				

arbitrarily set to 1.1, and those above that value were described as having a strong catalytic effect (Cr, U, Ba, Sn, Sr, Mn, Mg, Ca, Fe, and Mo, in that order). Those species having less but some effect were Co, Pb, Zn, Th, Na, Cu, and La, while V, Ni, Ga, and Cl (included for comparison purposes) had no effect at these volume fractions. The possibility of heterogeneous recombination on particles was admitted, but at these low volume fractions, the authors argued primarily for a homogeneous gas-phase mechanism involving H or OH reaction with the metal oxide or hydroxide (attributed to Jenkins).

$$H + MO + (X) = HMO + (X)$$

$$HMO + OH \text{ (or H)} = MO + H_2O \text{ (or } H_2\text{)},$$

in which M is a metal, and X is a third body.

In concurrent work with the alkaline earth metals, Cotton and Jenkins (1971) showed that Ca, Sr, and Ba added as a fine mist of a salt solution to premixed $H_2+N_2+O_2$ flat flames catalyzed the radical recombination for additive volume fractions in the range of 1 µL/L to 10 µL/L. By estimating the rates reaction in possible recombination mechanisms, they recommended the radical recombination mechanism to be:

$$MOH + H \Rightarrow MO + H_2$$
$$MO + H{_2}O(+X) \Rightarrow M(OH)_2(+X)$$
$$M(OH)_2 + H \leftrightarrow MOH + H_2O$$

Jenkins and co-workers extended their studies of metal-catalyzed radical recombination in premixed flames to study soot formation in diffusion flames. The strong effect of metals on soot formation in flames is well-known (Howard and Kausch, 1980). Cotton et al. (Cotton et al., 1971) added forty metals to co-flow diffusion flames of propane and N_2/O_2 mixtures and measured their effect on soot emissions and the smoke point. They postulated a mechanism for soot reduction whereby the catalytic reaction scheme listed above runs backwards, dissociating H_2O and H_2 into OH and H, which then oxidize the soot particles.

The question of heterogeneous versus homogeneous gas-phase chemistry was investigated by Bulewicz et al (Bulewicz et al., 1969). Through examination the variation of the intensity of emitted radiation as a function of wavelength from particles formed above a premixed $H_2+O_2+N_2$ flat flame with added spray of aqueous uranium salt, they determined that the particles (presumably uranium oxide) were up to 500 K above the gas temperature. They interpreted the particle temperature rise to be caused by the catalytic recombination of H and OH on the particle surfaces. Nonetheless, Tischer and Scheller (1970) pointed out that the spectral variation of the particle emissivity is unknown, and the gray-body assumption of Bulewicz is probably unjustified. They also argued that the excess temperature may be due to surface reactions other than radical recombination.

In similar work with premixed, fuel rich, flat flames of $H_2+O_2+N_2$, Jensen and Jones (1975) extended the classic $Li + H_2O \leftrightarrow LiOH + H$ photometric method to include the equilibrium for the reaction

$$Sr^+ + H_2O \leftrightarrow SrOH^+ + H,$$

in which $[SrOH^+]/[Sr^+]$ is measured mass-spectrometrically. Using both this new technique as well as the LiOH photometric method, they studied the catalytic flame radical recombination by tungsten and molybdenum, as well as once again confirming the strong effect of tin. Using flames at temperatures of 1800 K to 2150 K, and with metal addition at 1 µL/L to 110 µL/L, they collected data on the rates of radical recombination in the presence of the metal catalysts (added as tetramethyl tin, or hexacarbonyls of tungsten or molybdenum), and measured the major species present in flames inhibited by W and Mo. By analogy with the mechanisms they developed for calcium (Cotton and Jenkins, 1971)and iron (Jensen and Jones, 1974) in flames, they then postulated reaction mechanisms for W and Mo, and estimated the rates for the reactions in the catalytic cycles. For the conditions of their flames, the radical recombination cycles, for either W or Mo, were about five times faster then those of tin. (Using the results of Linteris et al., (2002) this translates to an effectiveness about ten times that of CF_3Br, or about one fifth that of $Fe(CO)_5$). For tungsten and molybdenum, the cycles are:

$$HWO_3 + H \leftrightarrow WO_3 + H_2$$
$$WO_3 + H_2O \leftrightarrow H_2WO_4$$
$$H_2WO_4 + H \leftrightarrow HWO_3 + H_2O$$
$$\overline{\qquad\qquad\qquad\qquad\qquad}$$
$$(net: H + H \leftrightarrow H_2)$$

and

$$HMoO_3 + H \leftrightarrow MoO_3 + H_2$$
$$MoO_3 + H_2O \leftrightarrow H_2MoO_4$$
$$H_2MoO_4 + H \leftrightarrow HMoO_3 + H_2O$$
$$\overline{\qquad\qquad\qquad\qquad\qquad}$$
$$(net: H + H \leftrightarrow H_2).$$

In continuing work, Jensen and Jones (1976) used similar techniques to study the radical recombination by cobalt added to a premixed, fuel rich, flat flame of H_2-N_2-O_2. With Co added as cyclopentadienylcobalt dicarbonyl at volume fractions of about 0.03 µL/L to 145 µL/L, and flame temperatures ranging from 1800 K to 2615 K, they spectroscopically identified the dominant cobalt-containing species to be Co, CoO, CoOH, and $Co(OH)_2$, with most of the cobalt being present in the flame as free Co atoms. Again by analogy with the Ca and Fe mechanisms, the Co mechanism was postulated to be:

$$Co + OH \Rightarrow CoO + H$$
$$CoOH + H \leftrightarrow Co + H_2$$
$$CoO + H_2O \leftrightarrow Co(OH)_2$$
$$Co(OH)_2 + H \leftrightarrow CoOH + H_2O$$
$$\overline{\qquad\qquad\qquad\qquad\qquad}$$
$$(net: H + H \leftrightarrow H_2),$$

with the first step added since Co is the dominant Co-containing species. The rates of these catalytic steps were again inferred from the experimental data. Cobalt appeared to be about 2/3 as effective as tin in these flames.

Hastie (1973a) also used mass spectrometry to study the effect of inhibitors on premixed flames. Using a Bunsen-type burner with a premixed CH_4-O_2-N_2 flame, he added $SbCl_3$ or $SbBr_3$ and detected the intermediate species through the flame. This work is described below.

Gas-phase Flame Retardants

Insight into mechanisms of metal flame inhibition can also be gleaned from studies of metal species added to materials as fire retardants (when their mode of action has been found to be in the gas phase). One such system is the antimony - halogen combination. Although they did not unravel the detailed mechanism, Fenimore and coworkers (Fenimore and Martin, 1966a; Fenimore and Jones, 1966; Fenimore and Martin, 1966b; Fenimore and Martin, 1972) showed that the relevant species act in the gas phase; they believed that the antimony moieties poisoned the flame, much as do brominated species. Similarly, Martin and Price (1968) observed that the addition of triphenylantimony to certain polymer substrates provided fire retardancy, even in the absence of halogen, and believed the mechanism involved antimony species in the gas phase. In a series of detailed experiments, Hastie and co-workers determined the mechanism of flame inhibition in the antimony-halogen system. Using a Knudsen effusion cell containing Sb_2O_3 over which passed HCl, they showed that $SbCl_3$ was evolved through a series of halogenation steps involving successive oxychloride phases (1973b). Using a molecular beam mass spectrometer, they studied both the pyrolysis products of polyethylene retarded by antimony-oxide/halogen, as well as intermediate species profiles in premixed $CH_4/O_2/N_2$ flames with added $SbCl_3$ and $SbBr_3$ (1975). For the pyrolysis studies, the major effused species from the polymer were $SbCl_3$ and SbOCl. In the flame studies (Hastie, 1973a), they found that $SbCl_3$ reacts readily to SbOCl, which then reacts with H to form SbO. They measured the major intermediate species of $SbBr_3$ flame inhibition to be SbO, Sb, HBr, and Br. They also measured the decrease in the hydrogen atom volume fraction with addition of $SbBr_3$, and developed a reaction sequence for the formation of the intermediate species, as well as for the gas-phase inhibition reactions. They argued strongly that the inhibition effect of antimony-oxide/halogen system is predominantly from the reaction sequence:

$$SbO + H \rightarrow SbOH*$$
$$SbOH + H \rightarrow SbO + H_2$$

net: $H + H \rightarrow H_2$

with a smaller effect from the usual bromine sequence, and an even smaller contribution from the equivalent chlorine cycle.

Ignition Studies

The effect of metals on ignition has been studied in both shock tubes and flames. Morrison and Scheller (1972) investigated the effect of twenty flame inhibitors on the ignition of hydrocarbon mixtures by hot wires, and found that $SnCl_4$ was the most effective inhibitor tested for increasing the ignition temperature; whereas the powerful flame inhibitors CrO_2Cl_2 and $Fe(CO)_5$ had no effect on the ignition temperature. Dolan and Dempster (1955) studied the effect of small particles (5 µm to 10 µm diameter) of metal compounds on suppressing the spark ignition of premixed natural gas-air mixtures in a vertical 7 cm diameter tube. They found that barium hydroxide octahydrate, barium chloride, and copper acetate monohydrate were each about three times less effective than sodium bicarbonate particles, while cuprous oxide and cobaltous chloride hexahydrate were each about 20 % less effective than $NaHCO_3$. Several studies have shown that metals can actually speed the ignition process in some chemical systems. In shock-tube studies, Matsuda and co-workers (1971) found that $Cr(CO)_6$ *reduced* the initiation time for reaction of CO or C_2H_2 with O_2 in shock-heated gases. The metal carbonyl was present at 0 µL/L to 50 µL/L, so the presence of any particles was precluded. The mode of action was believed to be gas-phase reactions involving CrO, CrO_2, and CrO_3. In later shock-tube work with $Fe(CO)_5$ in mixtures of $CO-O_2-Ar$, Matsuda (1972) found that a volume fraction of $Fe(CO)_5$ of few hundred µL/L greatly accelerated the consumption of CO_2. They postulated the effect to be due to oxidation of CO by metal oxides via: $FeO + CO \Rightarrow Fe + CO_2$, and noted that these reactions may be of importance since the mixtures were quite dry ($X_{OH} \approx 5$ µL/L). Interestingly, such accelerating oxidation pathways were also found to be important in the dry reaction of CO with N_2O in flames (Linteris et al., 2000b). In recent shock-tube studies of CH_4-O_2-Ar mixtures, Park et al. (2002) found that with 500, 1000, or 2000 µL/L of $Fe(CO)_5$, the ignition time was shorter than without the additive, again indicating a promotion effect of the metal additive; however, they did not determine the cause of the promotion for this moist system. Finally, in flash photolysis studies many years ago, Erhard and Norrish (1956) found that unlike $Pb(C_2H_5)_4$ and $Te(CH_3)_2$ which retarded hydrocarbon combustion, $Ni(CO)_4$, $Fe(CO)_5$, and $Cr(CO)_6$ all greatly accelerated the combustion.

Radical Recombination in Rocket Nozzles

Jensen and co-workers were motivated to study combustion inhibition by metal compounds largely by the desire to suppress afterburning in rocket nozzle exhausts. Using mechanisms developed for K, Ba, Fe, Mo, W, Cr, and Sn, Jensen and Webb (1976) calculated the amount of inhibitor required to suppress afterburning in the exhaust plume of a double-base propellant (unspecified composition, but usually a homogeneous mixture consisting of nitrocellulose and nitroglycerine). The products of the propellant reactions consist of a fuel rich mixture—principally CO and H_2, much like the recombination region above the fuel-rich premixed $H_2-O_2-N_2$ flames laboratory flames used to study the metal-catalyzed radical recombination. Their calculations (which do not include the effects of condensation to particles) indicate that W and Mo are about a factor of five less effective than Fe, Cr about a factor of six less effective, and Sn more than seven times less effective. These results are consistent with those of the flat flame measurements described above.

Particle formation was discussed by Jensen and Webb. Although they did not measure particles or calculate the degrading effects of particle formation on the suppression of afterburning in rocket motors, they did estimate the upper limit for radical recombination by the heterogeneous reactions on the particle surfaces. Their calculations indicated that although significant, the heterogeneous reactions could not suppress the afterburning, even for the smallest particle diameters assumed (10 nm). They also estimated that although the inhibiting species were probably volatilized in the combustion chamber, the characteristic times for condensation were probably of the same order as the residence time in the nozzle, indicating the potential for condensation. Their conclusions were that the metals held good promise for afterburning suppression in rocket motors, and that further work was necessary to estimate the rates of the catalytic cycle reactions and of the condensation rates of the metal derivatives at the conditions of their system.

Other Relevant Investigations of Metals in Combustion Systems

There have been a number of recent papers dealing with metals in flames. Crosley and co-workers (Westblom et al., 1994) added MMT to premixed flames with the purpose of studying its effect on NO formation. Adding MMT at volume fractions of about 0.5 µL/L to near-stoichiometric premixed low-pressure (522 Pa) propane-air flat flames in a McKenna burner, they measured the temperature and the relative concentrations of OH, H, O, CH, NO, and CO through the flame. While they observed no discernable effect of the MMT, this may have been due either to the low concentration of the additive, or the low pressure, both of which could limit the influence (Bonne et al., 1962; Linteris et al., 2002). They did, however, start to develop a mechanism for manganese which served as a basis for future efforts (Linteris et al., 2002).

Chromium reaction in a atmospheric pressure, premixed hydrogen-air flame flat flame was studied by Yu et al. (1998). Using microprobe gas sampling in the region downstream from the main reaction zone, they collected chromium species in the gas and condensed phase. The particle size distribution and the fraction of Cr as Cr(VI) was determined as a function of position. In addition, they assembled a kinetic mechanism for Cr reaction in flames through analogy with Boron and Aluminum combustion, and used the mechanism, together with the measured temperature profile, to calculate the fraction of hexavalent Cr in the downstream region from the flame zone. They also modeled the growth (but not the reaction) of the particle phase. For both particle growth and Cr(IV) formation, the calculations were able to predict the experimental trends. They concluded that further kinetic model development was necessary for accurate treatment of Cr speciation in flames. In later work, Kennedy et al. (1999) studied the morphology of the particles formed in a hydrogen–air–nitrogen co-flow diffusion flame with added chromium nitrate or chromium hexacarbonyl. They found that the morphology of the particles varied with the temperature of the flame and the source of the chromium.

Kellogg and Irikura (1999) performed theoretical calculations to predict the heats of formation as well as the enthalpies and free energies of reaction for the FeO_xH_y species thought to be important in iron inhibition. They found that nearly all of the reactions involving these species and potentially contributing to flame inhibition are exergonic at 1500 K. Hence, they suggested that further refinement of the inhibition mechanisms of iron would require knowledge of the actual kinetic rates of the inhibition reactions to improve upon the preliminary estimates of Rumminger et al. (1999).

In a comprehensive review of possible chemicals for use as halon alternatives, Tapscott et al. (2001) suggested that of the metals, Cu, Fe, Mn, and Sn, were worthy of further study. Since that time, premixed (Linteris et al., 2002) and co-flow diffusion flame (Linteris et al., 2004) studies have been performed for Fe, Mn, and Sn. No additional work has been reported for copper as a fire suppressant.

Detailed Studies with Iron

After the potential effectiveness of iron as a flame inhibitor was indicated (Walsh, 1954; Cheaney et al., 1959; Lask and Wagner, 1962), the behavior of iron pentacarbonyl was investigated in detail in several old investigations. The extraordinary flame inhibiting effectiveness of iron was first identified by Lask and Wagner (1962) in their screening study involving methane-, hexane-, and benzene-air premixed Bunsen-type flames with numerous additives. In continuing work, (Bonne et al., 1962) described the superior effectiveness of $Fe(CO)_5$ for reducing the burning velocity of hydrocarbon-air flames in nozzle burners. They found that an $Fe(CO)_5$ volume fraction of 100 µL/L reduces the burning velocity by 25 % when added to stoichiometric methane-air flames at atmospheric pressure. With oxygen as the oxidizer, or at reduced pressure, they found the effectiveness to be lowered. For hydrogen-air flames, $Fe(CO)_5$ was again much more effective than Br_2. They noted that at low volume fractions, the decrease in the burning velocity was proportional to the concentration of $Fe(CO)_5$, whereas for increasing concentrations of $Fe(CO)_5$, the relative influence seems to decrease. They postulated a homogeneous radical recombination mechanism at low volume fraction, and a heterogeneous one at higher.

To understand the detailed mechanism of $Fe(CO)_5$, Bonne et al. (1962) spectroscopically investigated premixed flat flames of methane with air or O_2. Unfortunately, at the low pressures for which the flame zone was expanded sufficiently to optically probe the flame (800 kPa), the kinetic effect of $Fe(CO)_5$ was very small. Nonetheless, for $Fe(CO)_5$ volume fractions up to 100 µL/L they observed that 1.) the peak OH volume fraction X_{OH} was unchanged, but shifted slightly downstream from the burner, and 2.) the decay rate of X_{OH} was increased in the presence of $Fe(CO)_5$, clearly indicating the effect of $Fe(CO)_5$ on radical concentrations in the flame. They measured FeO and Fe emission, as well as Fe absorption, and found that Fe and FeO emission peaked in the main high-temperature reaction zone, implicating these species in the radical recombination reactions. Although FeO had a double peak, with a minimum at the location where the rate of OH recombination was greatest, the authors did not feel that the decrease in FeO emission was correlated with a decrease in FeO concentration. They noted that solid particles were forming, that they were attempting to measure them, and said that the results of these measurements would be reported in the future (but no publications subsequently appeared).

In later work, the high effectiveness of iron was confirmed for methane-air premixed Bunsen-type flames, and extended to counterflow diffusion flames, by Reinelt and Linteris (1996). Their work indicated clearly that iron was very effective at low concentration, but that the effectiveness leveled off at higher concentration. Further studies with premixed flames of CO-O_2-N_2 and CO-N_2O-N_2 confirmed the effectiveness of $Fe(CO)_5$ in other systems.

Metals with Demonstrated Flame Inhibition Potential

From the above, we can assemble a list of metals with demonstrated inhibitory effects in flame systems. These are presented in Table 2. The metallic elements which have shown flame inhibition potential in experimental studies are listed down the left column, and the type of flame system used to determine the effectiveness is listed across the top. Under each category of flame system, the individual reference is listed.

Table 2 – Metals which have shown flame inhibiting properties. The type of experiment is shown at the top, followed by the specific reference. The elements are listed in the approximate order of demonstrated or expected effectiveness.

Element	Detailed Flame Studies					Engine Knock					Flame Screening Tests						Flat-Flame Radical Recombination						Ignition FR				
	Bonne (1962)	Reinelt and Linteris (1996)	(Linteris et al., 2002)	(Linteris et al., 2000a)	(Linteris et al., 2004)	Cheaney (Cheaney et al.,	Kovarik (Kovarik, 1994)	(Kuppu Rao and Prasad,	(Erhard and Norrish,	(Walsh 1954)	(Lask and Wagner, 1962)	(Vanpee and Shirodkar,	(Miller et al., 1963)	(Miller, 1969)	(deWitte et al., 1964)	(Rosser et al., 1963)	(Bulewicz and Padley,	(Bulewicz et al., 1971)	(Bulewicz and Padley,	(Jensen and Jones, 1976)	(Jensen and Jones, 1975)	(Jensen and Jones, 1974)	(Dolan and Dempster,	(Morrison and Scheller,	(Erhard and Norrish,	Hastie (1975).
Cr		X	X		X	X	X	X																		
Pb		X	X		X	X	X	X	X	X																
Fe	X	X	X	X	X	X		X	X	X	X		X		X											
Mn		X	X																							
Ni		X	X																							
W				X	X																					
Mo				X	X																					
Sn	X	X	X	X		X						X		X	X		X		X	X	X					
Co			X		X	X	X	X																		
Ti		X	X	X																						
Ge		X																								
Sb		X	X				X																			
Te	X	X	X																							
Tl	X																									
Bi	X																									
Cu		X				X	X	X	X																	
U			X	X																						
Zn			X																							
La			X																							
Th			X																							
Se	X																									

Flame Inhibition by Iron

Recent results provide convincing evidence that the strong flame inhibition by iron is due to a homogeneous gas-phase mechanism, which is described below. Following that, the experimental data showing iron's loss of effectiveness is described, along with a discussion of the parameters which most affect the particle formation which is believed to cause the loss of effectiveness.

Gas-Phase Mechanism

A detailed mechanism for the gas-phase iron-catalyzed radical recombination in flames was developed by Jensen and Jones (1974), and later, expanded by Rumminger et al. (1999). In both, the main catalytic cycle leading to radical recombination was postulated to be:

$$FeOH + H \leftrightarrow FeO + H_2$$
$$FeO + H_2O \leftrightarrow Fe(OH)_2$$
$$Fe(OH)_2 + H \leftrightarrow FeOH + H_2O$$
$$\overline{}$$
$$(\text{net: } H + H \leftrightarrow H_2).$$

This mechanism is shown schematically in Figure 2.

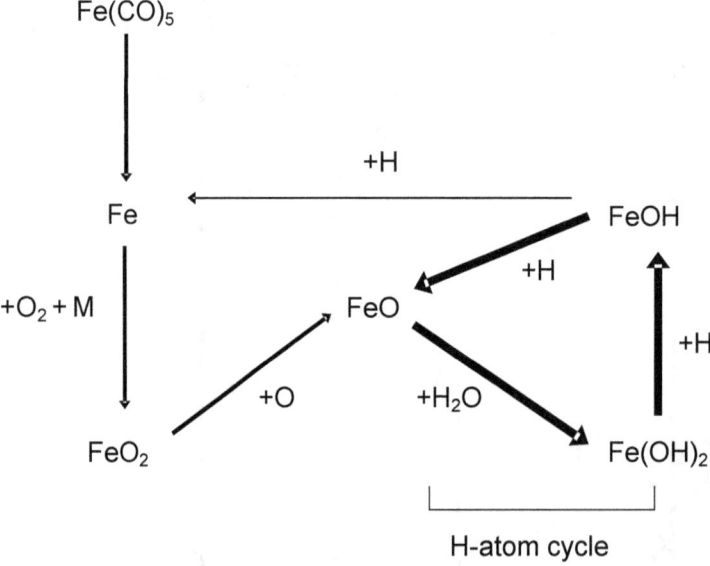

Figure 2 – Main catalytic radical recombination cycle of iron found to be important for methane-air flames.

Although the mechanism described in Rumminger et al. also included other catalytic cycles, they were not found to be particularly important in methane-air flames, premixed or diffusion (Rumminger and Linteris, 2000a). In work with premixed $CO-N_2-O_2$ flames, however, the following additional catalytic cycle was found to be much more important than the H-atom cycle:

$$Fe + O_2 + M \leftrightarrow FeO_2 + M$$
$$FeO_2 + O \leftrightarrow FeO + O_2$$
$$FeO + O \leftrightarrow Fe + O_2$$
$$\text{net: } O + O \leftrightarrow O_2 \; .$$

This new O-atom cycle together with the H-atom cycle are shown in Figure 3.

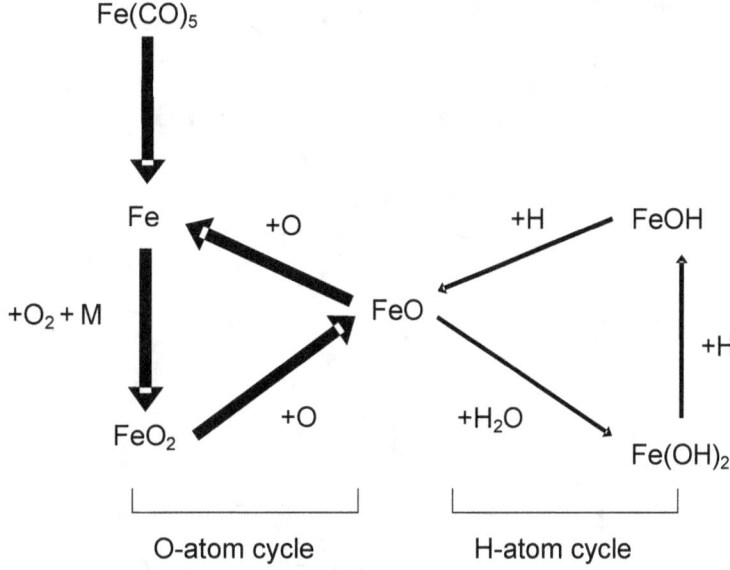

Figure 3 - Schematic diagram of radical recombination reaction pathways in found to be important for CO-H2-O2-N2 flames (Rumminger and Linteris, 2000b). Thicker arrows correspond to higher reaction flux. Reaction partners are listed next to each arrow.

Further, more recent computations of the thermochemistry of iron compounds at flame conditions support the possibility of many more radical recombination cycles (Kellogg and Irikura, 1999). In that study, seven iron species thought to exist at flame temperature were considered: Fe, FeH, FeO, FeOH, FeO_2, FeO(OH), $Fe(OH)_2$ (Rumminger et al., 1999), and the heat of reaction at 0 K, and the change in Gibbs free energy at 1500 K were calculated. Based on the results, however, very few of the considered reactions and cycles could be eliminated based on the thermodynamics. The complexity of the situation is illustrated in Figure 4, (from Kellogg and Irikura, (1999)), which shows schematically the possible inhibition cycles of iron. (Note that in the figure there are on the order of 50 possible cycles since [Fe] can be replaced by Fe, FeH, FeO, or FeOH.)

It is important to observe that even for iron, for which the most research has been performed, the mechanism is in very early stages of development. For example, the rates of the most important reaction steps in the mechanism were selected (within the uncertainty bounds suggested by Jensen and Jones (1974)) so as to provide the best agreement with experiments. The mechanism has been tested only for near stoichiometric premixed CH_4-N_2-O_2, CO-N_2-O_2, and CO-N_2O premixed flames, CH_4-O_2-N_2 counterflow diffusion flames, and CH_4-air cup-burner flames. Although at low Fe additive mole fraction the agreement was usually good, there were some conditions for which the predicted inhibition was off significantly. These cases include lean premixed CH_4-air flames (ϕ=0.9), CH_4-O_2-N_2 flames with an oxygen volume fraction in the oxidizer of 0.20, and the cup-burner flames with added CO_2 (at very low $Fe(CO)_5$ volume fraction).

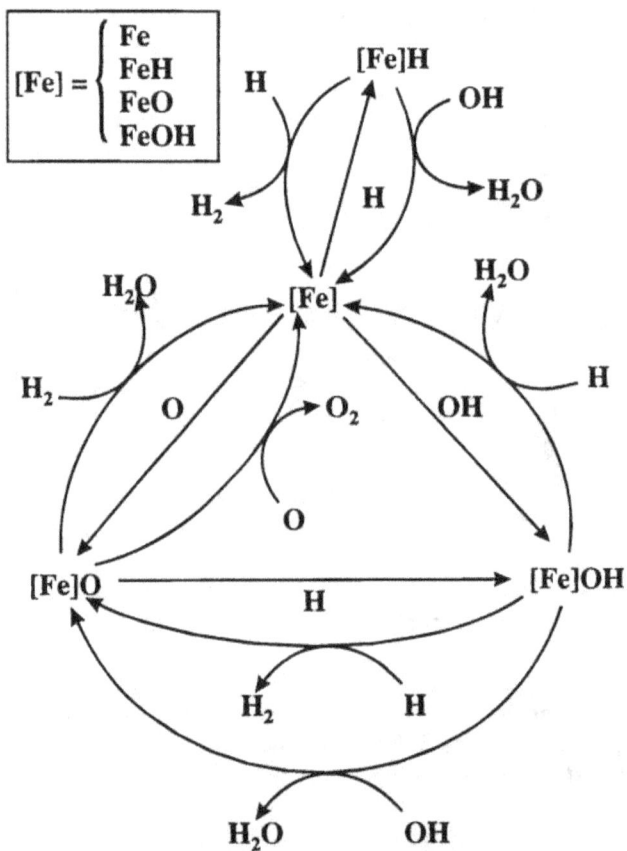

Figure 4 - Schematic representation of different classes of reactions which may contribute to iron's super-efficient flame suppression ability through the catalytic recombination of radical species from (Kellogg and Irikura, 1999).

The main ramification of the above gas-phase inhibition mechanism for fire suppression is that lowering radical concentrations lowers the overall reaction rate in the flame, weakening it. In laboratory flames, radical concentrations typically go above equilibrium levels, to super-equilibrium levels. This is particularly true for both premixed and high and moderate strain counterflow diffusion flames. At low strains, however, the radical super-equilibrium is less (see Figure 5). Radical concentrations in actual fires have not been measured. Nonetheless, recent research using cup-burner flames (Katta et al., 2003b) (which, to some extent, resemble small-scale fires) has shown that a stabilization region at the flame base, called the flame kernel, resembles a near-stoichiometric low-temperature premixed flame. Hence, the gas-phase *catalytic* radical recombination cycles should be as important in the stabilization of fires as they are for laboratory flames. This is partially confirmed since other gas-phase catalytic agents (for example, CF_3Br) are effective in both laboratory flames and full-scale fires.

One would expect that if the gas-phase inhibition mechanism of iron were the only consideration, it should be a very effective agent for extinguishing fires.

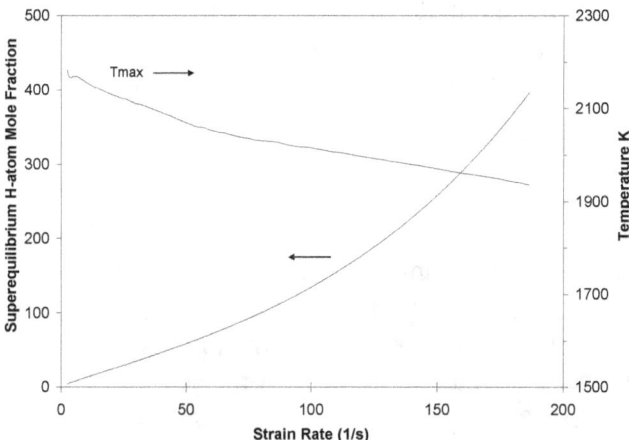

Figure 5 - **Calculated H-atom super-equilibrium ratio and peak temperature in counterflow diffusion flame as a function of strain rate.**

Nonetheless, recent work has shown that iron pentacarbonyl is much less effective in cup-burner flames than was expected based on the results in premixed or counterflow diffusion flames (Linteris et al., 2004). This was partially expected since earlier work with premixed flames (Rumminger and Linteris, 2000c) and counterflow diffusion flames (Rumminger and Linteris, 2002) indicated that particle formation would limit the effectiveness of iron to those situations in which it was added at high concentration. Nonetheless, even at low concentrations, $Fe(CO)_5$ was much less effective in cup-burner flames than was expected.

Iron's Loss of Effectiveness through Particle Formation

The effectiveness of iron as a flame inhibitor goes down as the mole fraction at which it is added goes up. This lower effectiveness was mentioned briefly by Jost and co-workers (1961), who said that the effectiveness of $Fe(CO)_5$ was lower at high mole fractions, and surmised that a particle inhibition mechanism may be at work at high concentration. They also said that work was underway to understand the role of particle formation; however, subsequent papers did not appear. Many other flame inhibitors also show lower effectiveness as their concentration increases (Noto et al., 1998). Nonetheless, the loss of effectiveness of iron is much more dramatic. Reinelt and Linteris (1996) quantified the loss of effectiveness of iron for premixed and diffusion flames. Rumminger et al. (1999) described a gas-phase kinetic mechanism for iron inhibition, and showed that the loss of effectiveness was not predicted by a gas-phase mechanism (ruling out low radical concentration as the reason for the loss of effectiveness). Linteris et al. (2002) have recently described the performance of iron, tin, and manganese in cup-burner flames. The loss-of-effectiveness of iron in each of these flame systems is described below.

It should also be noted that based on the encouraging results for the powerful flame inhibition properties of iron, extinction experiments of ferrocene together with an inert compound generated by a solid propellant gas generator (SPGG) were conducted in an enclosure containing a spray flame (Fallis et al., 2000). Unfortunately, the combination did not have the intended high efficiency. Although it was not possible to extract fundamental information concerning the lack of effectiveness for their flame configuration, the results of Holland and co-workers (2000) provide important evidence for a loss-of-effectiveness for iron, and motivate a search for an explanation.

Premixed Flames

For premixed flames, Figure 6 shows that the gas-phase mechanism predicts a continuing decrease in the normalized burning velocity ($S_L/S_{L,\ uninhibited}$) as [$Fe(CO)_5$] increases, whereas the experiments show a leveling off. Subsequent measurements of particles in the premixed flames showed that the loss of effectiveness of the iron was correlated with formation of particles (as illustrated in Figure 7 and Figure 8), and the scattering from the particles was correlated with the residence time in the flame (necessary for condensation to occur), as shown in Figure 9. In addition to measuring the particles with thermophoretic sampling and quantifying their size and agglomeration characteristics with transmission electron microscopy, they also calculated the maximum effect that the particles could have on the burning velocity. Constructing a "perfect heterogeneous inhibitor" model (in which any collision of a radical with a particle recombines the radical to a stable species), they showed that heterogeneous reactions cannot account for the measured flame speed reduction of $Fe(CO)_5$. Figure 10 shows the experimental data for normalized flame speed reduction by $Fe(CO)_5$ along with the prediction for the perfect gas-phase inhibitor and perfect heterogeneous inhibitor (for different assumed particle diameter sizes). As shown, the perfect gas-phase inhibitor mechanism shows burning velocity reductions fairly close to that of $Fe(CO)_5$; whereas, the perfect heterogeneous mechanism does not show enough inhibition, about a factor of eight too low, even for perfect collisions with 10 nm particles. Interestingly, the slope of the $Fe(CO)_5$ curve in Figure 10 in the "flattening out" region is of the same order as that of the perfect 10 nm diameter particles. Hence, Jost et al.'s (1961) postulate that the region of lower effectiveness may be due to radical recombination on particles surfaces, may be true. Note also that the slope in the flat region is of the same order as that for CF_3Br in these premixed flames.

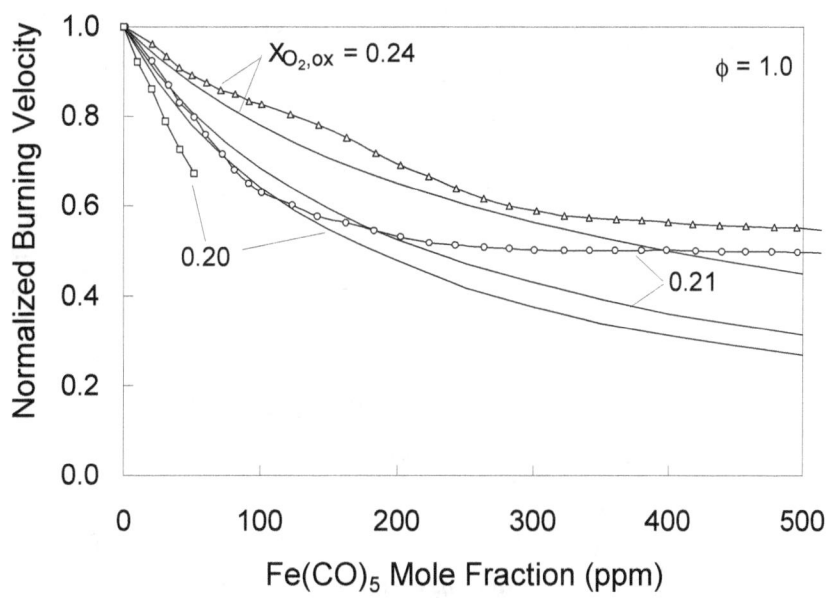

Figure 6 - Calculations and measurements of normalized burning velocity of premixed $CH_4/O_2/N_2$ flames with $X_{O_2,ox}$ = 0.20 (squares), 0.21 (circles), and 0.24 (triangles) for ϕ = 1.0 (source: (Rumminger et al., 1999))

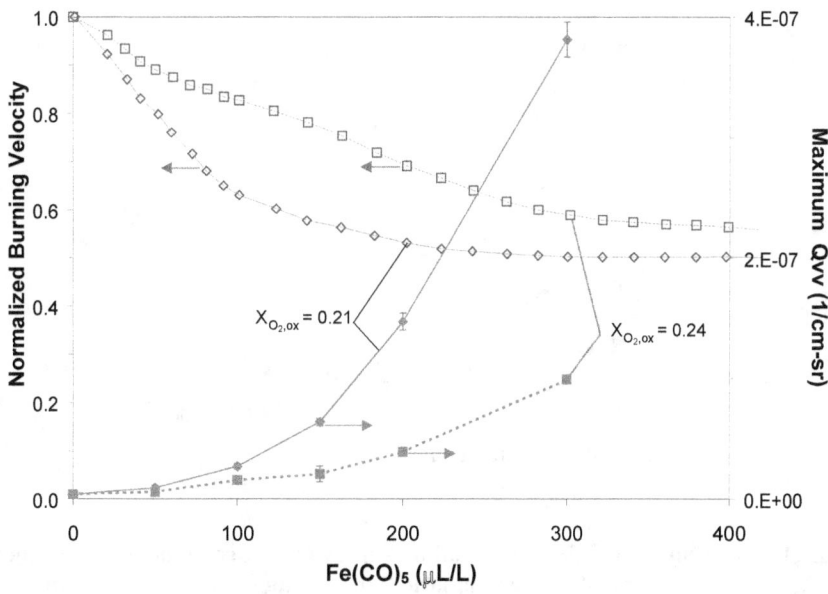

Figure 7: Normalized burning velocity (Reinelt and Linteris, 1996) and maximum measured scattering signal Q_{vv} (Rumminger and Linteris, 2000c) for ϕ=1.0 CH_4 flame with $X_{O_2,ox}$ = 0.21 and 0.24.

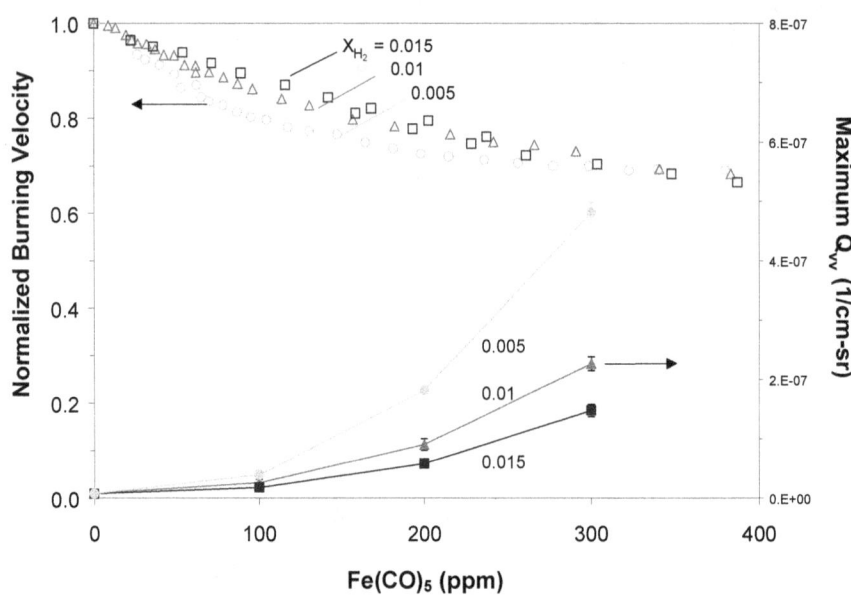

Figure 8: Maximum scattering signal and normalized burning velocity (Rumminger and Linteris, 2000b) for CO-H_2 flames as $Fe(CO)_5$ concentration varies (Rumminger and Linteris, 2000c).

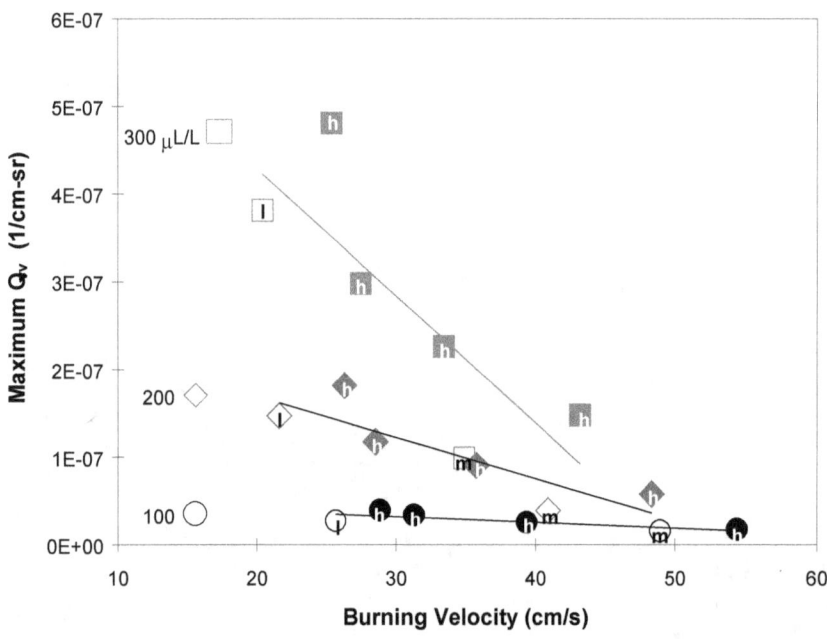

Figure 9 - Maximum Q_{vv} for flames of CH_4 (open symbols) and CO (closed symbols) as a function of the burning velocity. The letters correspond to the adiabatic flame temperature (low, medium, and high, 2220, 2350, and 2470 K), while the symbol shape (square, diamond, and circle) corresponds to the loading of $Fe(CO)_5$: (100, 200, and 300) μL/L (Rumminger and Linteris, 2000c).

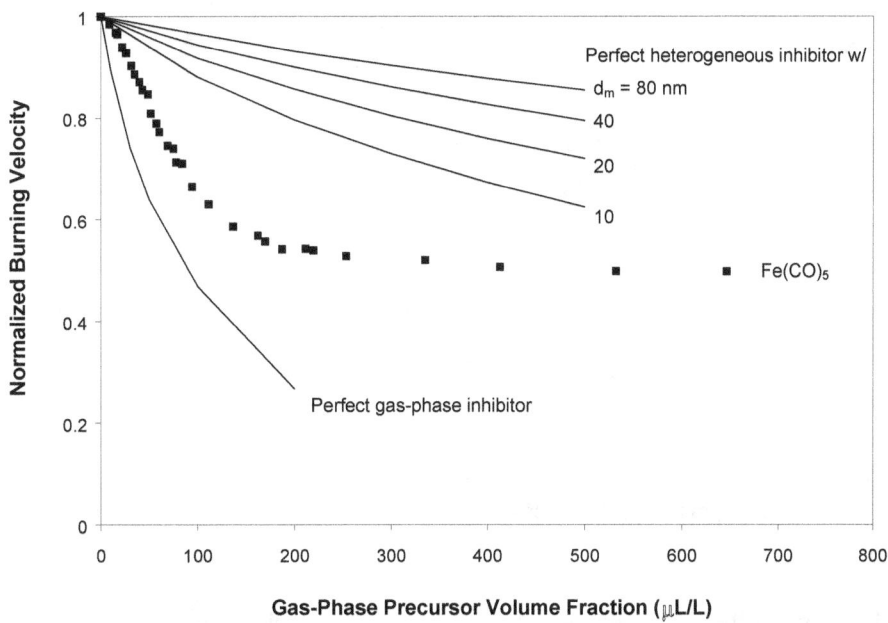

Figure 10 - Calculated normalized burning velocity for several diameters d_m of ideal heterogeneous inhibitor. Also shown are $Fe(CO)_5$ data (Reinelt and Linteris, 1996), and calculated normalized burning velocity using the perfect gas-phase inhibitor mechanism (Babushok et al., 1998; Rumminger and Linteris, 2000c).

Counterflow Diffusion Flames

Similar experimental and modeling results were described for counterflow diffusion flames. As shown in Figure 11, the gas-phase model predicts more inhibition at higher [$Fe(CO)_5$] than do the experiments. Measured particle scattering signals again showed that the loss of effectiveness (and discrepancy between the measured and gas-phase predicted inhibition) increased as particle scattering signal increased; i.e., particle formation was correlated with loss of effectiveness of the $Fe(CO)_5$. (Figure 12 It is important to note that in the counterflow diffusion flames, the propensity to form condensed-phase iron particles <u>in the flame region</u> was not the only mechanism shown to be correlated with loss of effectiveness of iron. If particles formed in one part of the flame, but were unable (due to entrainment or thermophoresis) to be transported to the reaction zone, the active iron species could be sequestered from the region in which they must be present to inhibit the flame. By adding the $Fe(CO)_5$ to either the fuel or air stream, and changing the flame location relative to the stagnation plane (by diluting the fuel or oxidizer with N_2), Rumminger et al. clearly showed the effect of drag and thermophoretic forces on reducing particle (and hence iron) transport to the region of radical chain branching.

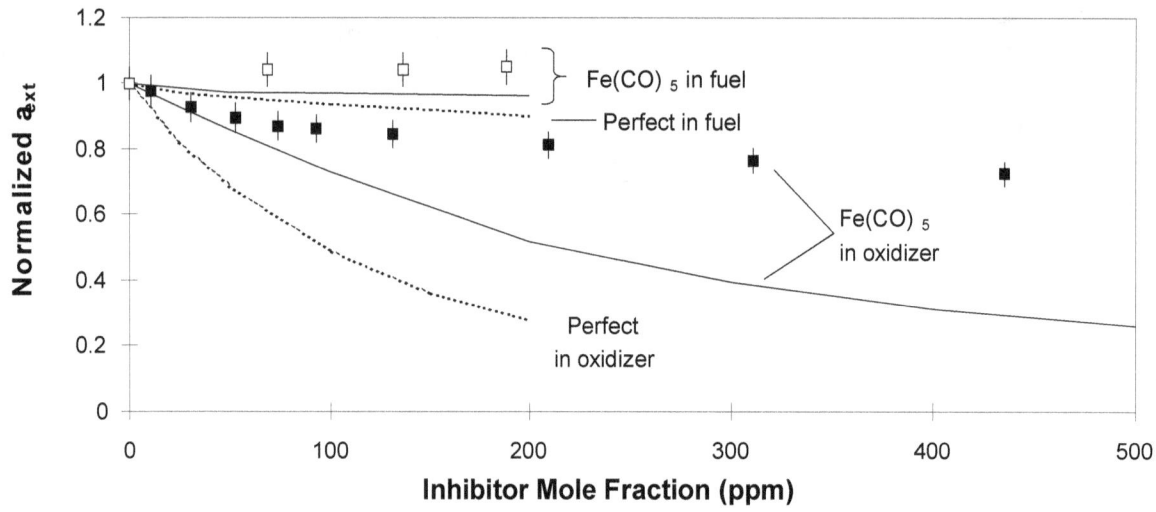

Figure 11 - Normalized extinction strain rate for counterflow diffusion flames. Closed symbols: measurements with the Fe(CO)$_5$ in the oxidizer; open symbols: measurements with Fe(CO)$_5$ in the fuel; solid lines: calculations with Fe(CO)$_5$; dashed lines: calculation with perfect inhibitor (source: (Rumminger and Linteris, 2000a)).

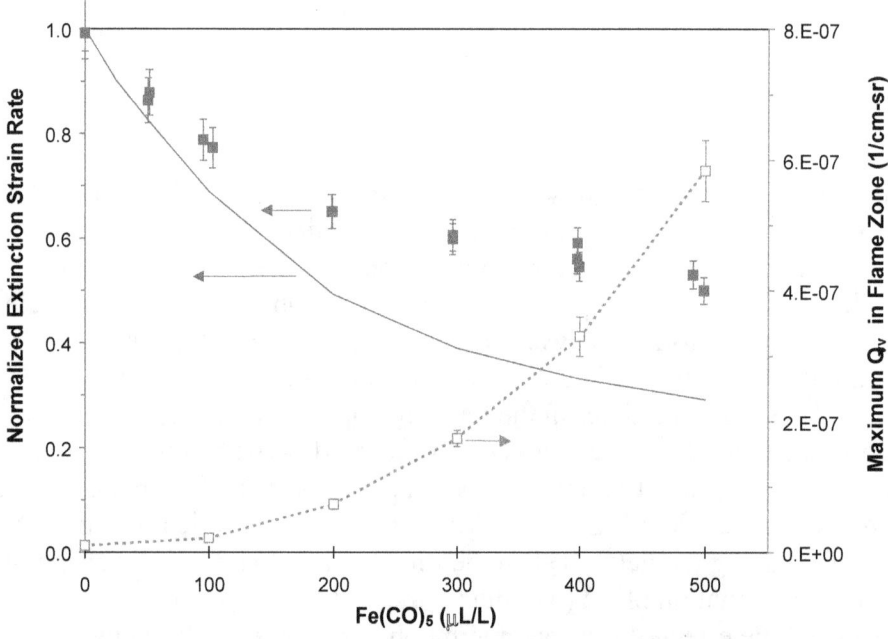

Figure 12 - Correlation between inhibition effect in counterflow diffusion flames and maximum measured scattering signal Q_{vv}. Filled points are experimental normalized a_{ext}, solid line is calculated a_{ext} (Rumminger and Linteris, 2000a). Open symbols connected by dotted lines are maximum measured Q_{vv}. Particle data collected at 75 % of a_{ext} (Rumminger and Linteris, 2002).

Cup Burner Flames

Linteris et al. have recently tested $Fe(CO)_5$, TMT, and MMT in cup-burner flames (Linteris and Chelliah, 2001; Linteris et al., 2002; Linteris et al., 2004). Based on the loss of effectiveness at higher $Fe(CO)_5$ concentrations that was demonstrated in premixed and diffusion flames, it was not expected that iron pentacarbonyl alone would be effective in cup-burner flames. Notwithstanding, there was much suggestion in the literature that a combination of a good catalytic agent and an inert agent would prove to be an effective combination. In this case, the overall reaction rate is lowered in part through radical recombination by the catalytic agent, and in part through the lower temperature caused by the added diluent. This approach has been discussed in work since the 1950s (Rosser et al., 1959; Rosser et al., 1963; Lippincott and Tobin, 1953; Lott et al., 1996; Noto et al., 1998; Rumminger and Linteris, 2000b) which suggested that combinations of thermally acting and catalytic agents might prove beneficial. To test these suggestions, Linteris et al. added $Fe(CO)_5$, MMT, or TMT to a cup burner of methane and air, and measured the amount of CO_2 required for extinction (Linteris et al., 2004). This approach is conceptually the same as the classic oxygen index test used for assessing material flammability (Fenimore and Jones, 1966). In that test, the oxygen volume fraction in the air stream at blowoff (i.e., the oxygen index) is determined for solid, liquid, or gaseous fuels with chemical additives in either the fuel or oxidizer.

Unfortunately, the effectiveness of CO_2 combined with any of the metal agents was much less than anticipated. Figure 13 shows the measured CO_2 volume fraction required for extinction as a function of the catalytic agent volume fraction in the air stream. Data are presented for Br_2, CF_3Br, $Fe(CO)_5$, MMT, and TMT. As shown, although the metals are still more effective than CF_3Br at low concentration, they are not nearly as effective as expected from the results in premixed and counterflow diffusion flames. The reasons for this loss of effectiveness are described below.

In order to understand the lower effectiveness of metals in cup-burner flames compared to premixed and counterflow diffusion flames, several steps were taken. First, particle measurements in the cup-burner flames inhibited by $Fe(CO)_5$ were made. The results (shown in Figure 14 to Figure 17) indicate that particles are present both inside and outside (but not coincident with) the luminous flame zone, and that higher $Fe(CO)_5$ loadings produced higher particle scattering signals. In order to understand the particle formation and chemical inhibition, numerical modeling of the cup-burner flames inhibited by $Fe(CO)_5$ were performed (Linteris et al., 2004), using the gas-phase only numerical model developed previously. This model has predicted the blow-off condition of methane and air cup-burner flames with added CO_2 (Katta et al., 2003b) and CF_3H (Katta et al., 2003a). The temperature field and the velocity vectors for methane-air cup-burner flames with 10 % CO_2, and 0 and 100 µL/L of $Fe(CO)_5$ are shown in Figure 18, while Figure 19 shows the calculated blow-off behavior. As in other flame configurations, the loss of effectiveness of iron and a discrepancy between predicted (gas-phase model) and measured effectiveness were both correlated with the formation of particles (see Figure 20). To understand the propensity for particle formation, the degree of super-saturation

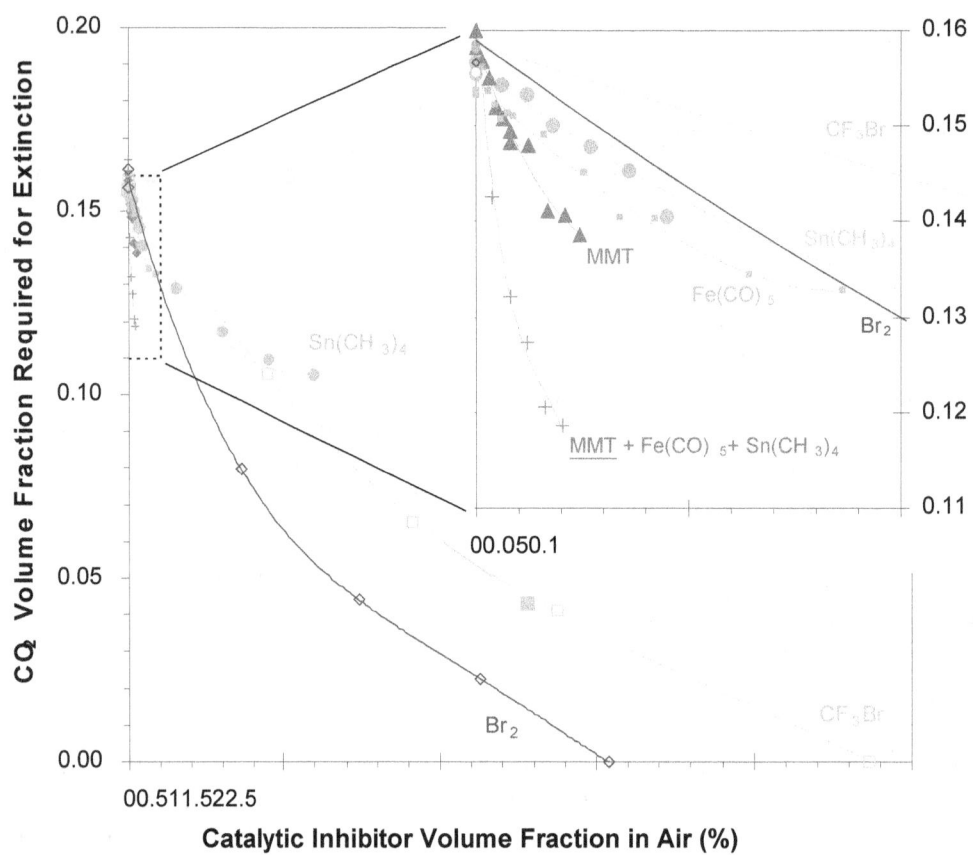

Figure 13 – Volume fraction of CO_2 required for extinction ($X_{CO_2,ext}$) of methane-air cup burner flames as a function of the volume fraction of catalytic inhibitor added to the air stream. Inset shows region in dotted box with expanded scales (Linteris et al., 2004).

of some of the iron-containing intermediates was calculated through the flame, using the detailed flame structure obtained from the model, together with vapor pressure data available from the literature. For a height above the cup burner which passes through the flame kernel (i.e., the stabilization region), Figure 21 shows the radial profile of temperature and volume fractions of iron species and radicals, as well as the supersaturation ratio (which is the ratio of the calculated species partial pressure to the planar vapor pressure at the local conditions). The supersaturation ratio is highest for FeO, followed by Fe and Fe(OH)$_2$, and the values decrease as the radial location of peak temperature is approached. Note that vapor pressure data for FeOH, FEOOH, and FeO$_2$ are not available, so their condensation potential has not been assessed. The condensation potential is strong since the temperature of the flame kernel is much lower than the relevant regions of premixed or counterflow diffusion flames.

Finally, the numerical model was extended to include calculation of the particle trajectory for inert particles added to the flame, including the effects of gravity, drag, and thermophoretic forces. This was done since early estimates (Linteris et al., 2002) were that thermophoresis may have been driving the particles away from the flame region. The results of the calculations (Figure 22) show that near the flame base, there is some deviation of the particles both up and down around the reaction kernel; however, examination of the estimated radial and axial thermophoretic velocities shows them to be much less than the gas velocity. Consequently, the particles still pass directly into the reaction kernel, so the effect of thermophoresis near this region is expected to be minor. Nonetheless, the other results described above provide evidence that the loss of effectiveness is due to particle formation.

A review of the results from previous work with particle formation in premixed, counterflow diffusion, and cup-burner flames inhibited by $Fe(CO)_5$ outlined the importance of the following physical effects with respect to effective chemical inhibition:

1.) gas-phase transport of the active iron-containing species to the region of high H-atom concentration is necessary for efficient inhibition.
2.) Particle formation near the location of peak [H] can act as a sink for the iron-containing intermediate species and reduce the catalytic effect.
3.) The volume fraction of inhibitor influences condensation since at low values, it may be below its saturation value.
4.) The available residence time affects particle growth.
5.) If the particles are small enough, they can re-evaporate upon passing into the high-temperature region of the flame.
6.) Thermophoretic forces can be large in the flame and re-distribute particles away from peak [H].
7.) Convection and drag forces combined with the existing flow field in the flame can prevent particles from reaching the region of peak [H].
8.) The flame temperature of the stabilization region of cup-burner flames is much lower than in premixed or counterflow diffusion flames, exacerbating the condensation potential.

In order to assess the condensation potential of other flame inhibiting metals, it is necessary to know their concentrations in the flame, as well as their local vapor pressure. The availability of these data for the metals listed in Table 2 is discussed below.

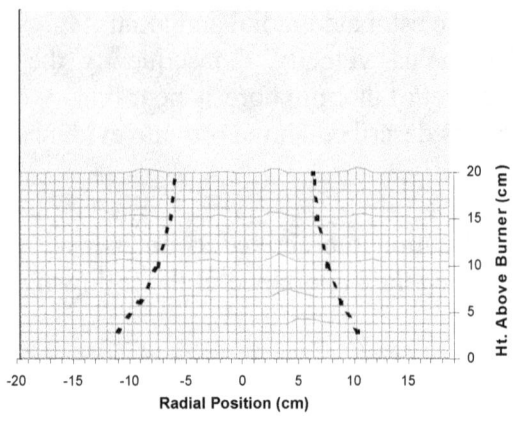

Figure 14 to Figure 17 – Scattering cross section for laser light at 488 nm as a function of radial position and height above burner in methane-air cup-burner flame with 8 % CO_2 and $Fe(CO)_5$ in air at specified volume fraction. Dotted lines show flame location from a digitized video image of the uninhibited flame (Linteris et al., 2004).

Figure 14 - $Fe(CO)_5$ in air at 100 μL/L.

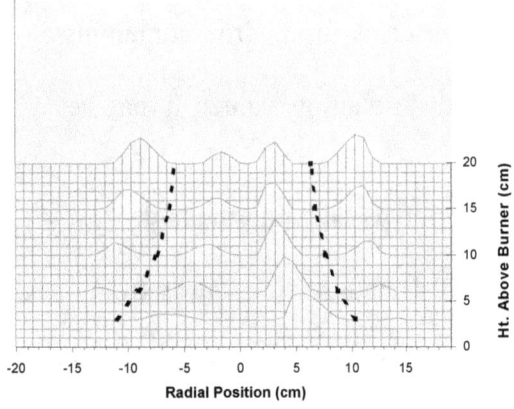

Figure 15 – $Fe(CO)_5$ in air at 200 μL/L.

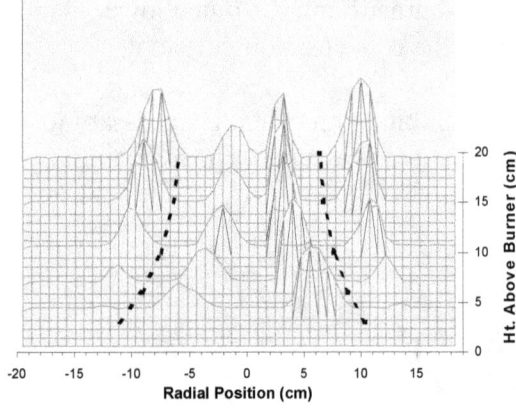

Figure 16 – $Fe(CO)_5$ in air at 325 μL/L.

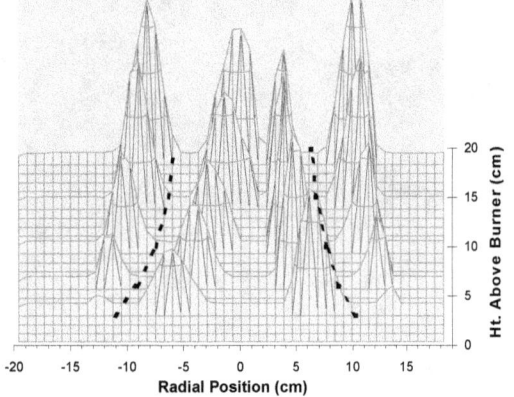

Figure 17 – $Fe(CO)_5$ in air at 450 μL/L.

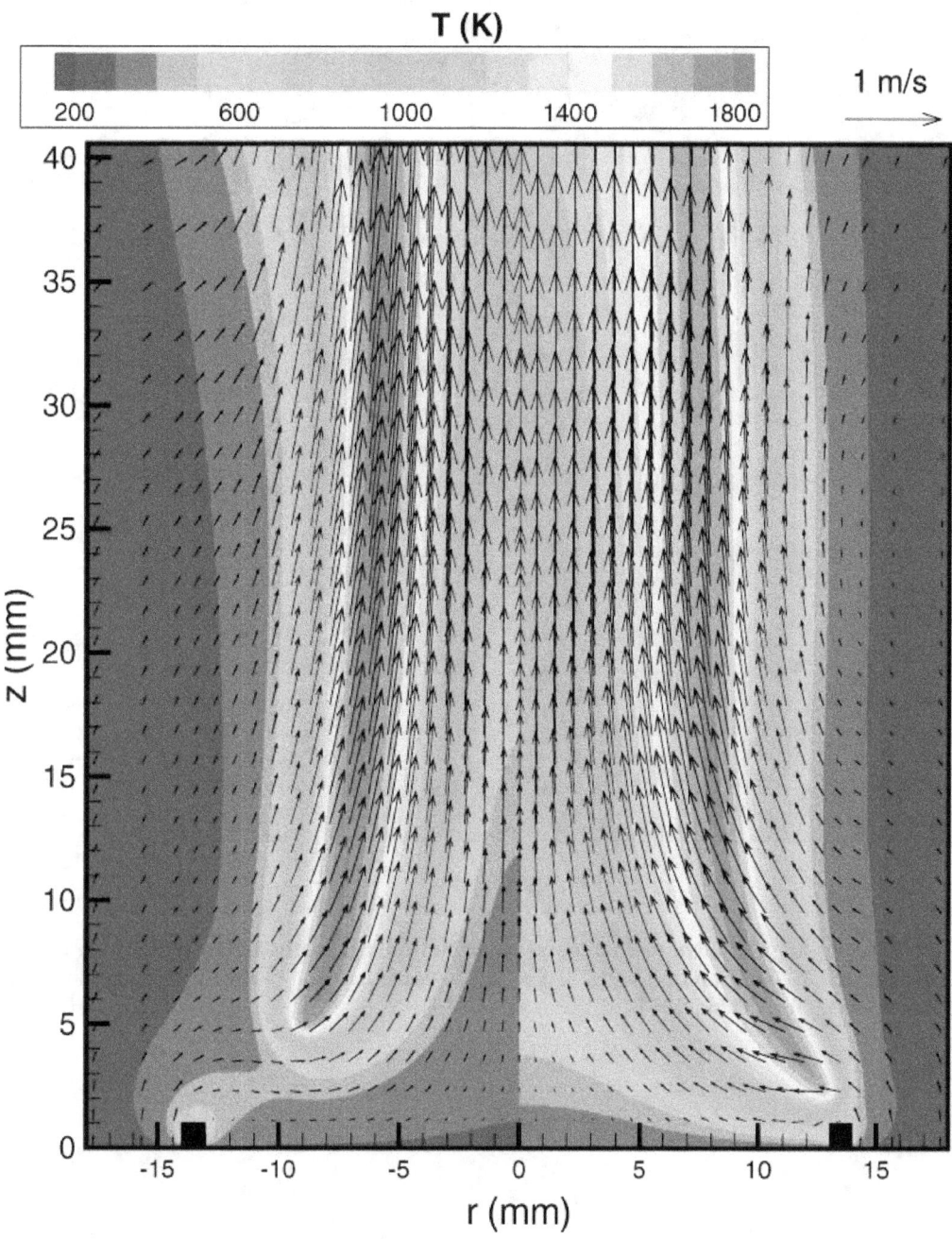

Figure 18 – Calculated temperature (color scale) and velocity vectors (arrows) for methane-air cup-burner flame with an oxidizer stream CO_2 volume fraction of 10 %, with (left) and without (right) an added $Fe(CO)_5$ volume fraction of 100 µL/L (Linteris et al., 2004).

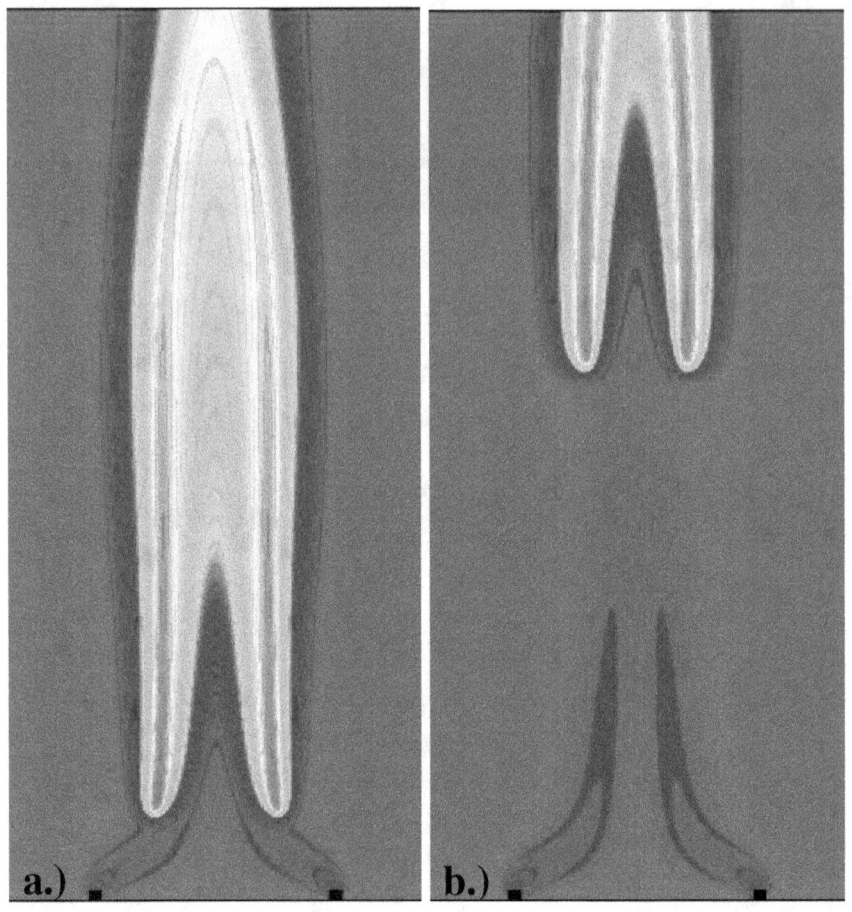

Figure 19 – Two-D color map of calculated temperature in cup-burner methane-air flames with 10 % CO_2 in the oxidizer stream, and a.) 0.011 and b.) 0.012 % $Fe(CO)_5$ volume fraction in the air stream, illustrating the blowoff phenomenon (Linteris et al., 2004).

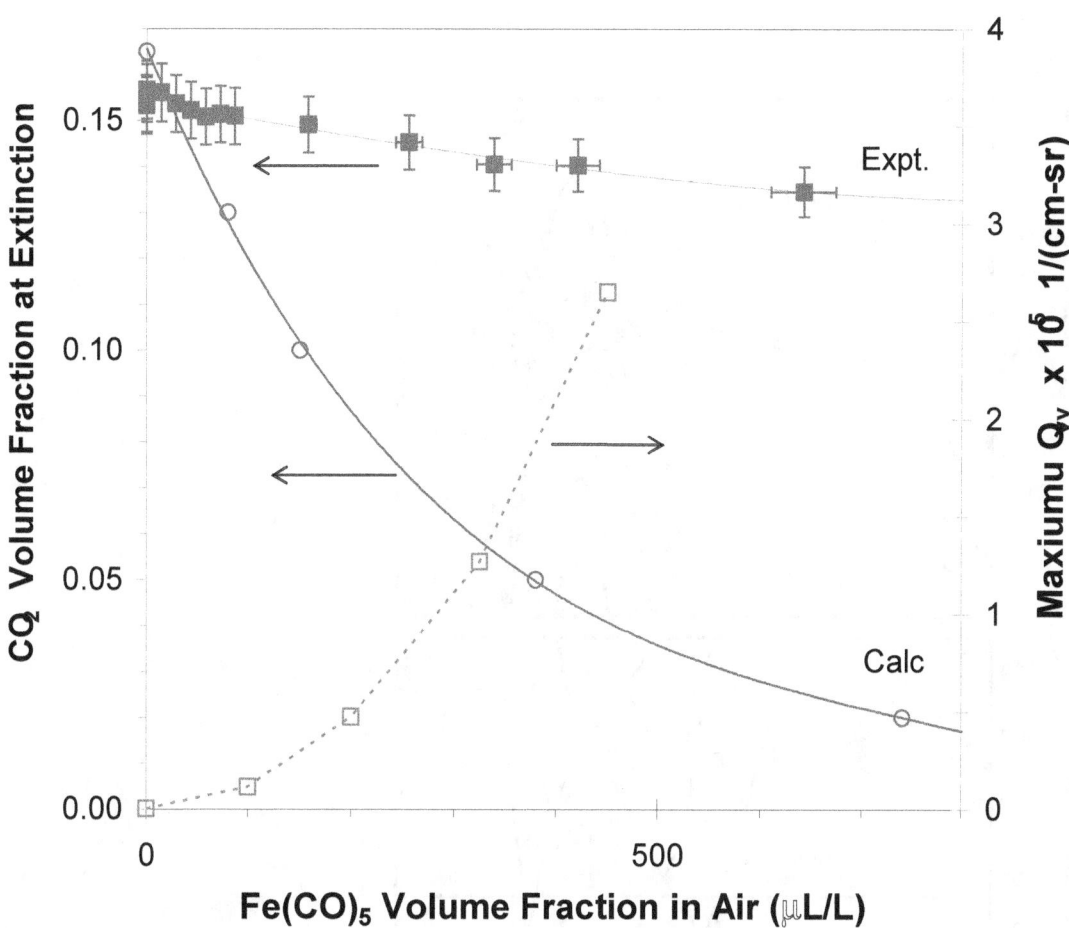

Figure 20 – Experimental and numerically predicted extinction volume fraction of CO_2 (left axis) and peak measured scattering cross section (right axis), as a function of the volume fraction of $Fe(CO)_5$ in the air stream; from (Linteris and Chelliah, 2001) and (Linteris et al., 2004).

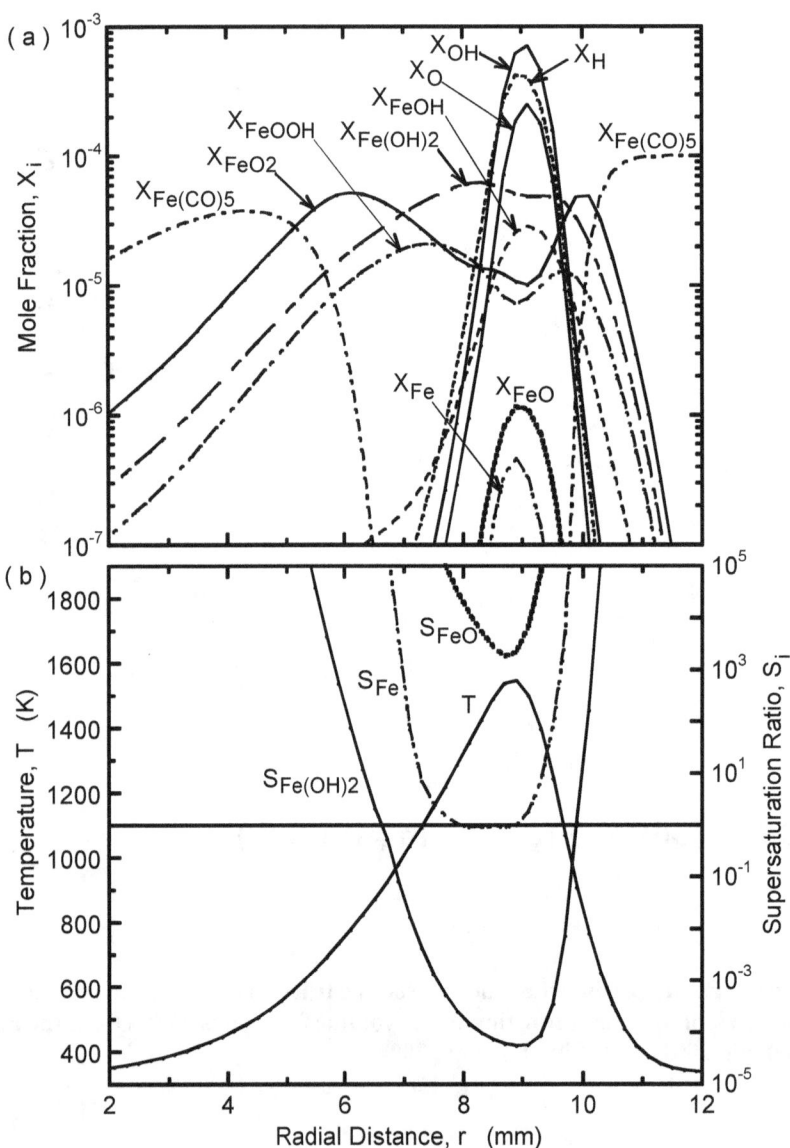

Figure 21 – a) Calculated iron-containing and major species volume fraction X_i as a function of radial position at the height above the burner of 4.8 mm (corresponding to the location of the reaction kernel in the flame base); and b) the super-saturation ratio, S_i, for Fe, FeO, and $Fe(OH)_2$ (Linteris et al., 2004).

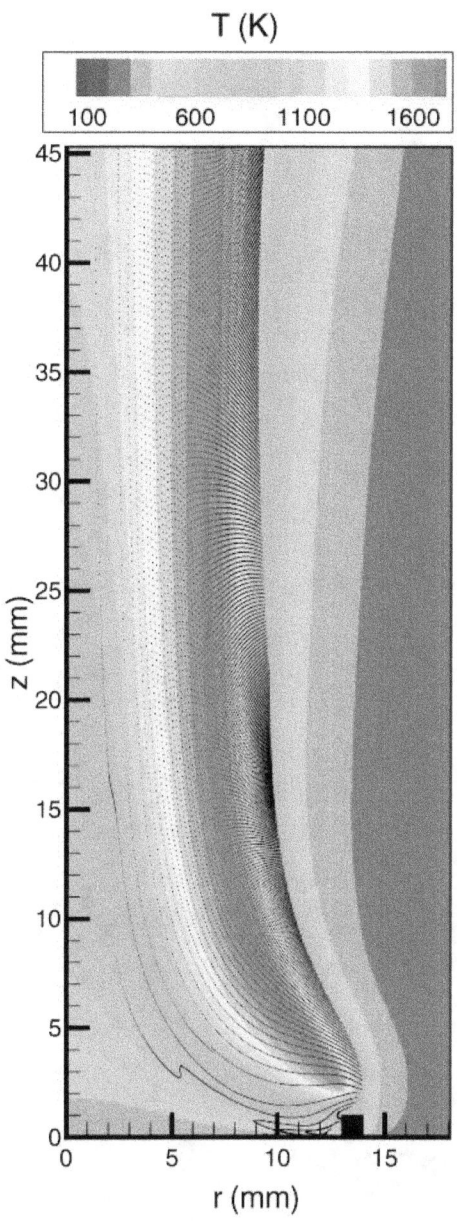

Figure 22 – Calculated particle trajectories for free-molecular-regime particles in a CH4 – air flame with 10 % CO_2 in the oxidizer stream (Linteris et al., 2004).

Equivalent Flame Inhibiting Species of Other Metals

In general terms, there are two basic approaches for identifying the equivalent flame inhibiting species for other metals besides iron. One is through analogy with existing known mechanisms (Cotton and Jenkins, 1971; Bulewicz et al., 1971; Jensen and Jones, 1974; Jensen and Webb, 1976; Jensen and Jones, 1976; Jensen and Jones, 1975), and the other is through a comprehensive consideration of all possible radical recombination cycles based on estimated thermodynamic properties (Kellogg and Irikura, 1999). Both depend upon generation of a list of probable species in the flame, which usually comes from spectroscopic identification of the species in flame systems. The latter generates a large number of possible species and reactions, which must then be culled down based on measured reaction rates of the possible inhibition reactions. Each of these approaches can be a research project in itself, for each metal inhibitor. It should also be noted that even for iron (for which the most research has been completed) there exist major limitations in the mechanistic description. For example, only a sub-set of the possible catalytic cycles has been included in the inhibition mechanism; the rates of the key reactions have been adjusted to provide agreement with recent burning rate data; some experimental results are not predicted well by the mechanism, and the vapor pressure of some key intermediates is not available. Nonetheless, the mechanism developed represents good progress in understanding flame inhibition by these super-effective moieties, and provides a basis for further work. The extension of the work for iron to other metals is a challenging but tractable task. As a first step, the following sections describe what it known, for other metals, about the species present in flames, the inhibition mechanisms, and the possibility for loss of effectiveness due to condensation. Discussion is provided for Cr, Pb, Mn, W, Mo, Sn, Co, Cu, and Sb. These metals have been selected because they have the most indicated potential for flame inhibition; they are listed in decreasing order of the expected inhibition efficiency (based on existing experimental results).

Cr

Although chromium as a flame inhibitor has been the subject of several studies (Lask and Wagner, 1962; Miller, 1969; Bulewicz and Padley, 1971b; Vanpee and Shirodkar, 1979), and it is believed to be one of the most powerful catalytic metals for radical recombination, very little is known about its mechanism. As described above, Bulewicz and Padley (1971a) did a detailed study in which chromium carbonyl or chromium salts were added to a flat premixed rich H_2-O_2-N_2 flame. They detected the presence of particles, and estimated the upper limit on the radical recombination on the particle surfaces to be of the same order as the un-catalyzed radical recombination rate. They proposed that the following balanced reactions are important in chromium inhibition, but did not develop an explicit mechanism.

$$Cr + OH \leftrightarrow CrO + H$$
$$CrO + OH \leftrightarrow CrO_2 + H$$

$$CrO_2 + H_2O \leftrightarrow HCrO_3 + H$$

Further, they speculated that at very low additive concentrations, the following gas-phase reactions might be important.

$$CrO + H + X => CrOH^* + X$$
$$CrOH^* + H => CrO + H_2$$

Pb

Despite it's known effectiveness as a flame inhibitor (see Table 2), there is very little detailed knowledge of the homogeneous gas-phase catalytic mechanism of lead. The active species were detected to be Pb and PbO in flash photolysis studies (Erhard and Norrish, 1956).

Mn

The mechanism for Mn is believed to be very similar to that of iron (Linteris et al., 2002). Mn reacts with O_2 to form MnO_2, which reacts primarily with radicals to form MnO. The catalytic radical recombination cycle consists of:

$$MnO + H_2O \leftrightarrow Mn(OH)_2$$
$$Mn(OH)_2 + H \leftrightarrow MnOH + H_2O$$
$$MnOH + H \text{ (or OH, O)} \leftrightarrow MnO + H_2 \text{ (or } H_2O, OH)$$

$$\text{net: } H + H \leftrightarrow H_2 \text{ (or: } H + OH \leftrightarrow H_2O)$$

Although flame equilibrium calculations show that the species MnH is present at relatively large concentrations, the contribution of reactions of this species to the inhibition effect is relatively small.

W

Jensen and Jones (1975) found the dominant species in tungsten inhibition to be WO_3, HWO_3, and H_2WO_4. They developed the following mechanism for tungsten inhibition by analogy with their mechanisms for Ca and Fe.

$$HWO_3 + H \leftrightarrow WO_3 + H_2$$
$$WO_3 + H_2O \leftrightarrow H_2WO_4$$
$$H_2WO_4 + H \leftrightarrow HWO_3 + H_2O$$

$$\text{(net: } H + H \leftrightarrow H_2)$$

From their experiments with premixed flat flames, they estimated the rate of the cycle to be about ten times that of CF_3Br, or about one fifth that of iron.

Mo

Using the premixed of H_2-N_2-O_2 flat flame, Jensen and Jones (1975) found the dominant species in molybdenum inhibition to be MoO_3, $HMoO_3$, and H_2MoO_4, and they developed the following inhibition mechanism :

$$HMoO_3 + H \leftrightarrow MoO_3 + H_2$$
$$MoO_3 + H_2O \leftrightarrow H_2MoO_4$$
$$H_2MoO_4 + H \leftrightarrow HMoO_3 + H_2O$$

$$(\text{net: } H + H \leftrightarrow H_2).$$

Sn

A mechanism for flame inhibition by tin was developed by Babushok and co-workers (Linteris et al., 2002). The tin atom formed as a result of TMT decomposition quickly reacts with O_2 through the reactions $Sn + O_2 (+M) \leftrightarrow SnO_2$, and $Sn + O_2 \leftrightarrow SnO + O$. The former reaction leads to SnO from the reaction of SnO_2 with CO, H, or other radicals. Conversely, the latter reaction forms SnO directly, and is fast at room temperature as compared to the analogous reaction of iron atom. Formation of SnO leads to the following reactions with H and HCO radicals:

$$SnO + H + M \leftrightarrow SnOH + M$$
$$SnO + HCO \leftrightarrow SnOH + CO$$

which, together with the radical scavenging reactions of SnOH, complete the catalytic radical recombination cycle of tin:

$$SnOH + H \leftrightarrow SnO + H_2$$
$$SnOH + OH \leftrightarrow SnO + H_2O$$
$$SnOH + CH_3 \leftrightarrow SnO + CH_4$$
$$SnOH + O \leftrightarrow SnO + OH.$$

The net effect of the dominant inhibition reactions can be shown as:

$$SnO + H + M \leftrightarrow SnOH + M$$
$$SnOH + H \leftrightarrow SnO + H_2$$

$$\text{net: } H + H \leftrightarrow H_2$$

Equilibrium calculations show that SnO is the major tin species in the products of a stoichiometric methane–air flame with added TMT.

Co

Cobalt was added to a premixed, fuel rich, flat flame of H_2-N_2-O_2. by Jensen and Jones (1976) who studied the radical recombination rate in the post combustion region having temperatures ranging from 1800 K to 2615 K. The dominant cobalt containing species were found to be Co, CoO, CoOH, and $Co(OH)_2$, with most of the cobalt being present in the flame as free Co atoms. By analogy with the Ca and Fe mechanisms, the Co mechanism was postulated to be:

$$Co + OH \Rightarrow CoO + H$$
$$CoOH + H \leftrightarrow Co + H_2$$
$$CoO + H_2O \leftrightarrow Co(OH)_2$$
$$Co(OH)_2 + H \leftrightarrow CoOH + H_2O$$
$$\text{-----------------------------------}$$
$$(\text{net: } H + H \leftrightarrow H_2),$$

in which the first step was included since Co is the dominant Co-containing species. The rates of these elementary steps were estimated so as to provide agreement with the experimental radical decay rate. On this basis, cobalt appeared to be about 2/3 as effective as tin in these flames.

Cu

Little has been reported on the actual flame inhibition mechanism of copper. The copper-containing species present above a Meker burner supplied with H_2-N_2-O_2 mixtures with added copper salts were found to be Cu, CuO, CuOH, and CuH (Bulewicz and Sugden, 1956b; Bulewicz and Sugden, 1956a). Their concentrations were related by the balanced reactions:

$$Cu + H_2O \leftrightarrow CuOH + H$$
$$Cu + OH + X \leftrightarrow CuOH + X$$
$$Cu + H_2 \leftrightarrow CuH + H$$
$$Cu + H + X \leftrightarrow CuH + X$$

Sb

Hastie (1973a) used premixed flames with mass spectrometry to measure intermediate species in antimony inhibition. The only antimony species he identified in the flame were Sb, SbO, and halogenated products of the additive ($SbCl_3$ or $SbBr_3$). He postulated the mechanism to be:

$$SbO + H \rightarrow SbOH*$$
$$SbOH + H \rightarrow SbO + H_2$$

Potential for Loss of Effectiveness of Other Metals

There exist several approaches for estimating whether other metals will have similar loss of effectiveness to that of iron. One approach is to have experimental data for flame systems in which the loss of effectiveness is evident. This requires that the inhibitor be added at volume fractions high enough to allow the loss of effectiveness (which is believed to be due to condensation of the active gas-phase species to particles). For example, many of the early studies with metal compounds did not provide data to high enough volume fractions to show the loss of effectiveness (Lask and Wagner, 1962) or the inhibiting effect was not presented as a function of additive volume fraction (so the decreasing effectiveness was not illustrated) (Vanpee and Shirodkar, 1979; Miller et al., 1963; Miller, 1969; Bulewicz and Padley, 1971a; Bulewicz et al., 1971; Jensen and Jones, 1975; Jensen and Jones, 1976). Another clue that a loss of effectiveness may occur is a reported presence of particles in some flame system. Although the presence of particles will depend upon the temperature of the flame, the concentration at which the metal moiety is added, and the residence time for particle formation, the observed presence of particles in one flame system is an indication that it may be important in other flame systems as well. Finally, the potential for condensation can be assessed by considering the local metal species volume fraction as compared to the its local vapor pressure in the flame. A limitation of this method is that it relies upon knowledge of the metal species present in a flame system, the mechanism of inhibition, as well as the vapor pressure (or gas-phase and condensed-phase thermodynamic data). Often, this information is incomplete. Further, the kinetic rates of the formation of more stable oxides of the metal must be known to assess the contribution of those compounds to condensed-phase particles (since often, the vapor pressure of these oxides is very low; e.g., Fe_2O_3). Below, the available information related to the potential for condensation is presented for each metal species of interest.

Cr

Bulewicz and Padley (1971b) found particles in the post combustion region of premixed flat flames inhibited by chromium through visual observation, and by detecting continuum radiation from them. They also observed saturation in both the inhibiting effect and in the concentration of gas-phase Cr in the flame. In addition, solid deposits were observed on the burner in premixed flames inhibited by CrO_2Cl_2, indicating the potential for particle formation (similar deposits have been observed for premixed flames inhibited by Sn, and Mn (Linteris et al., 2002) and for Fe (Reinelt and Linteris, 1996). Hence, it seems clear that the potential for condensation of Cr species exists and may be important in other flames systems. To further examine this potential, the vapor pressure of Cr, CrO, CrO_2, and CrO_3 above Cr_2O_3, under neutral and oxidizing conditions is shown in Figure 23. Note that since Cr_2O_3 is a particularly stable oxide, these vapor pressures will be lower than if the equilibrium included other, less stable condensed-phase oxides. For example, in Figure 23, the blue curve for Cr is data for Cr vapor above Cr solid. Nonetheless, since it is not yet know for *any* metal inhibitor what the composition of the condensed phase oxide actually is, the potential for condensation of Cr species is illustrated. That is, many of the chromium oxides appear to have quite low vapor pressures.

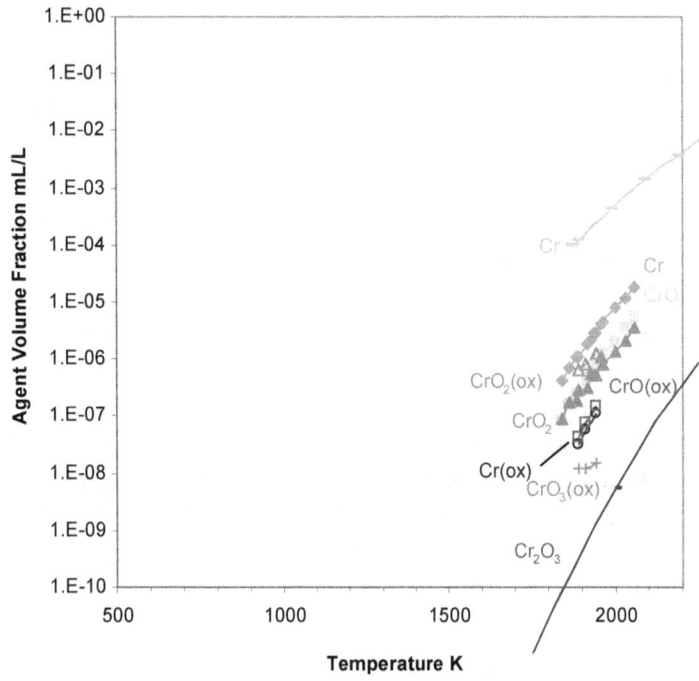

Figure 23 – Volume fraction of Cr, CrO, CrO$_2$, and CrO$_3$ above Cr$_2$O$_3$ at 10133 Pa under neutral (closed symbols) and oxidizing (open symbols) conditions above Cr$_2$O$_3$ (reddish hues), and Cr above itself (blue); from (Nesmeianov, 1963; Grmiley et al., 1961; Gurvich et al., 1993).

Pb

The presence of PbO particles was detected in many engine knock studies (Cheaney et al., 1959; Kovarik, 1994; Kuppu Rao and Prasad, 1972; Walsh, 1954), and it was often argued that the strong inhibition of knock occurred from heterogeneous radical recombination on the particle surfaces. Although lead has been shown to be one of the more effective metal agents in flame screening studies (Lask and Wagner, 1962), the additive mole fraction (maximum 150 µL/L) was not increased high enough to show the loss of effectiveness that might occur. In the flash photolysis studies of Norrish et al. (1956), however, the TEL concentration was systematically increased, and a drop-off in the effectiveness was observed above a certain volume fraction. Hence, condensation is likely to limit lead effectiveness for at least some applications. It should also be noted, that the high effectiveness of Pb in premixed flames cannot be from particles since, as shown by Rumminger et al. (Rumminger and Linteris, 2000c), the collision rate of radicals with solid particles is not high enough to account for the observed inhibition of premixed flames by super-effective agents such as iron or lead. The vapor pressure of Pb and PbO are shown as a function of temperature in Figure 24. The data for PbO$_2$ are too low to appear on this figure, and hence, condensation is clearly possible for Pb-inhibited flames if PbO$_2$ readily forms in the inhibition cycle, or if other higher oxides readily form.

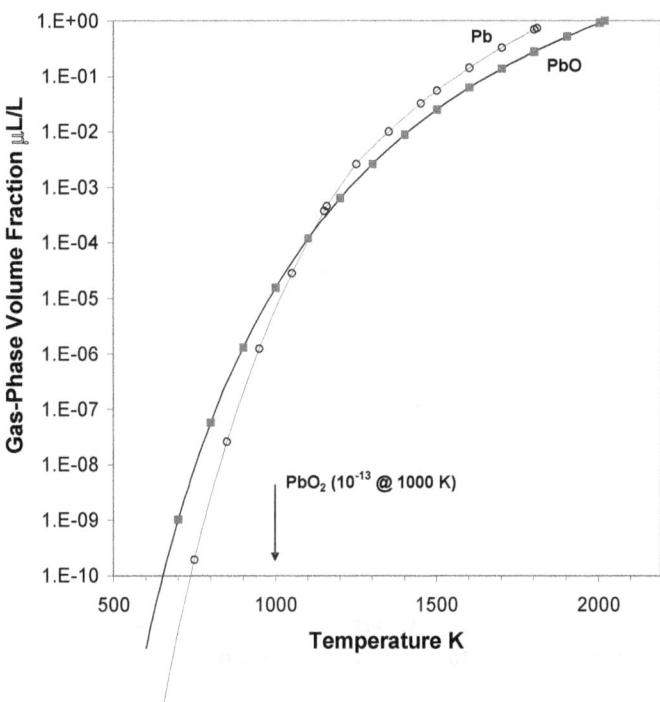

Figure 24 - Pb species gas-phase volume fraction at equilibrium over the condensed-phase (at 1 atm), i.e., vapor pressure, Pb and PbO from (Barin and Knacke, 1973), PbO$_2$ from (Gurvich et al., 1993; Barin et al., 1977).

Mn

The effect of MMT on the flame speed of methane-air mixtures was measured by Linteris et al. (2002), as shown in Figure 25. The data clearly show a saturation effect, which was beyond that due to lowering of the radical concentrations. Additional tests in cup-burner flames showed a much lower effectiveness than expected, as well as a rapid loss of effectiveness for MMT as its volume fraction was increased (see Figure 13 inset, above). The presence of solid particles was also observed visually as were solid deposits on the burner. Finally, the vapor pressure of Mn, MnO, and MnO$_2$ are shown below in Figure 26. Since the volume fraction necessary for flame inhibition is in excess of 200 µL/L, the potential for condensation is apparent.

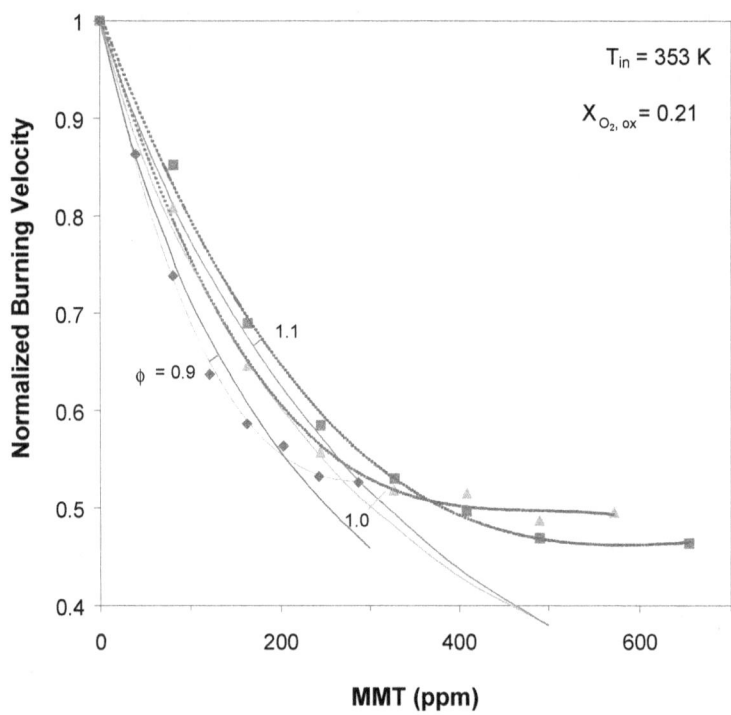

Figure 25 – Normalized burning velocity of premixed $CH_4/O_2/N_2$ flames inhibited by MMT with $X_{O2,ox}=0.21$ and $\phi=0.9$, 1.0, and 1.1 (dotted lines: curve fits to data; solid lines: numerical predictions; from: (Linteris et al., 2002)).

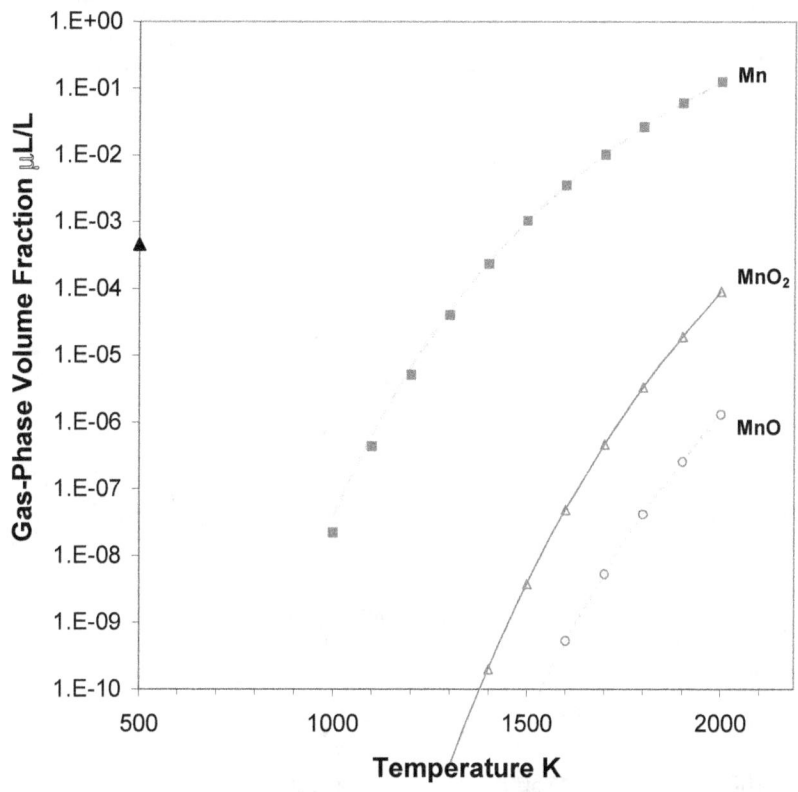

Figure 26 – Mn species gas-phase volume fraction at equilibrium over the condensed-phase (at 1 atm), i.e., vapor pressure, Mn, MnO, and MnO_2 from (Gurvich et al., 1993).

W

There are no data in the literature concerning particle formation or loss of effectiveness of flames inhibited by molybdenum compounds. A search for vapor pressure or thermodynamic data for molybdenum oxides, hydrides, and hydroxides allowed generation of the results in Figure 27. As shown, WO_3 has quite high vapor pressure at flame temperatures; on the other hand, the vapor pressures of WO_2 and W are low enough to allow particle formation. Depending upon the flame reactions of W-containing compounds and their rates, as well as the vapor pressures of other compounds which form in flames, condensation may or may not be important.

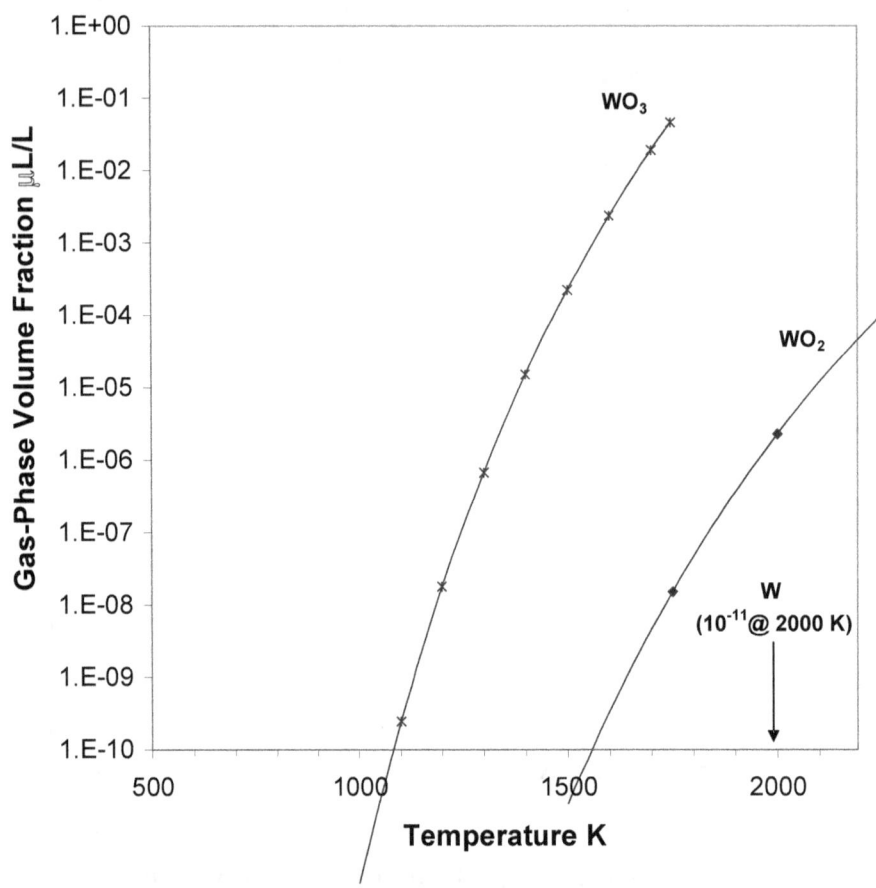

Figure 27 - W species gas-phase volume fraction at equilibrium over the condensed-phase (at 1 atm), i.e., vapor pressure, W, WO_2, and WO_3, from (Nesmeianov, 1963; Gurvich et al., 1993; Stull, 1947).

Mo

There are likewise no data in the literature concerning particle formation or loss-of-effectiveness for flames inhibited by Mo-containing species. The vapor pressure of Mo, MoO_2, and MoO_3 for flame conditions are shown in Figure 28. As for W compounds, the only species thought to participate in the catalytic cycle for which vapor pressure data are available (WO_3 and MoO_3) have quite high vapor pressures.

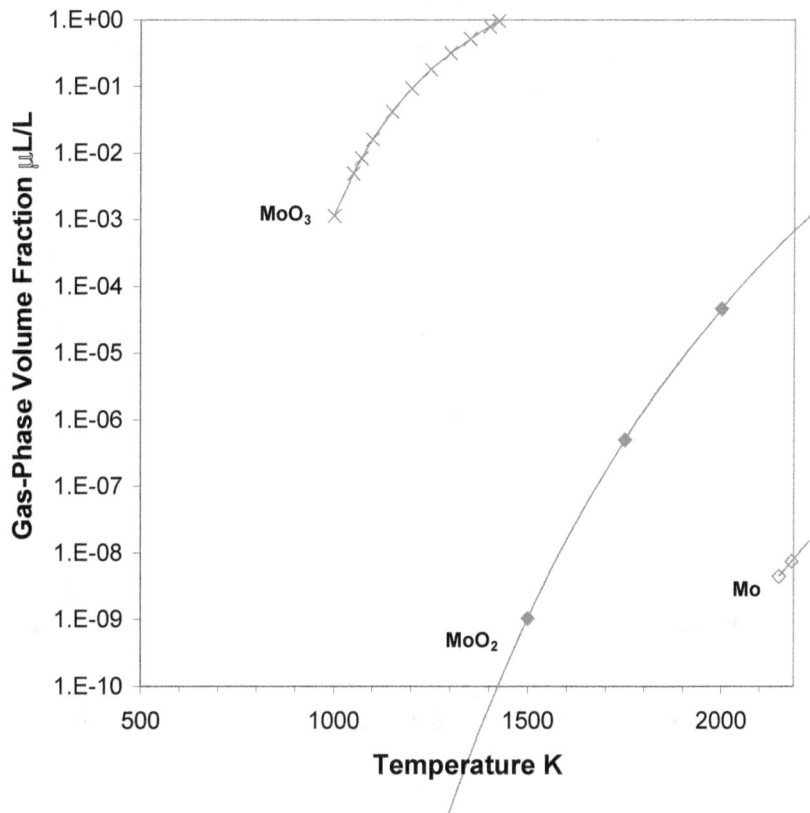

Figure 28 - Mo species gas-phase volume fraction at equilibrium over the condensed-phase (at 1 atm), i.e., vapor pressure, Mo, MoO_2, and MoO_3, from (Nesmeianov, 1963; Gurvich et al., 1993; Stull, 1947).

Sn

The results for Sn are similar to that of Mn. At the volume fractions necessary for effective flame inhibition, a loss of effectiveness was observed in premixed flames, although not as dramatic as for Mn (see Figure 29 below); particle deposits were observed on the burner. Cup-burner tests showed much lower effectiveness than expected, as well as a further loss of effectiveness as the Sn volume fraction increased (see Figure 13 inset, above), and solid particles were observed both in the flame and deposited on the burner. The vapor pressure of Sn, SnO, and SnO_2 is shown in Figure 30. Clearly, condensation is possible at the 3000 µL/L volume fraction needed for flame inhibition.

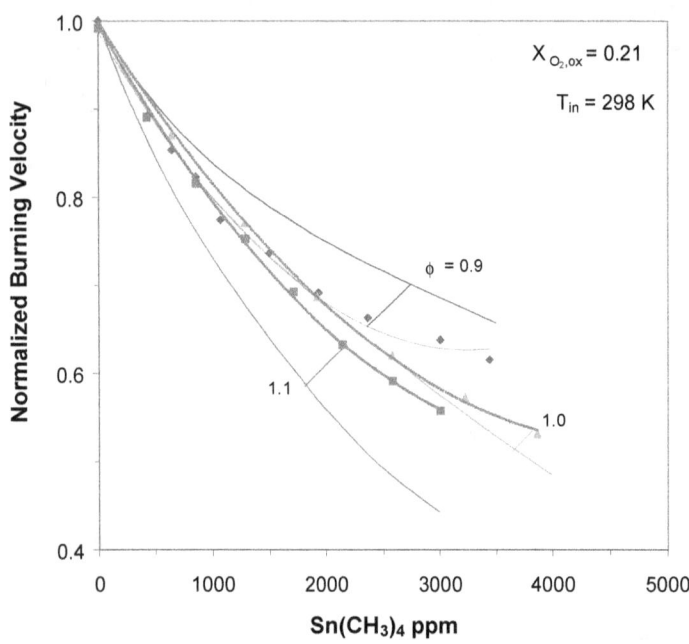

Figure 29 - Normalized burning velocity of premixed $CH_4/O_2/N_2$ flames inhibited by TMT with $X_{O_2,ox}=0.21$ and $\phi=0.9$, 1.0, and 1.1 (dotted lines: curve fits to data; solid lines: numerical predictions; from (Linteris et al., 2002)).

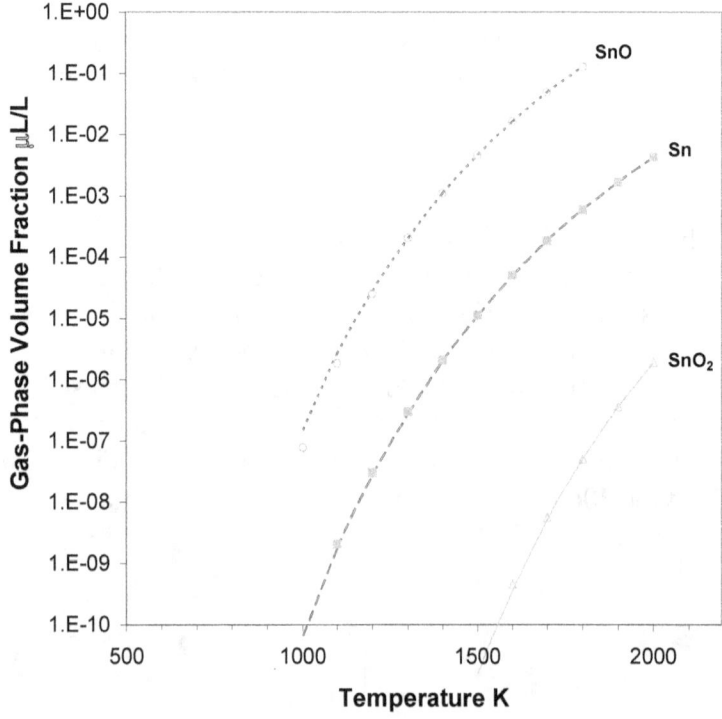

Figure 30 – Sn species gas-phase volume fraction at equilibrium over the condensed-phase (at 1 atm), i.e., vapor pressure, Sn, SnO, and SnO_2 from (Gurvich et al., 1993).

Co

The data for Co effectiveness in various flame systems has not be presented as a function of Co concentration. Hence, loss of effectiveness at higher concentrations has not been demonstrated. Likewise, particles formation in flames with added Co has also not been discussed. The available vapor pressure data for Co oxides and hydroxides is shown below. Since CoO is an intermediate species in the proposed inhibition mechanism of Co in H_2-O_2-N_2 flames, and total Co loading in the flame is expected to be on the order of several thousand µL/L, condensation may occur. On the other hand, depending upon the relative rates of the inhibition reactions, the volume fraction of CoO may not build up too high, and hence it's propensity to condense could be lower than Figure 31 might lead on to believe.

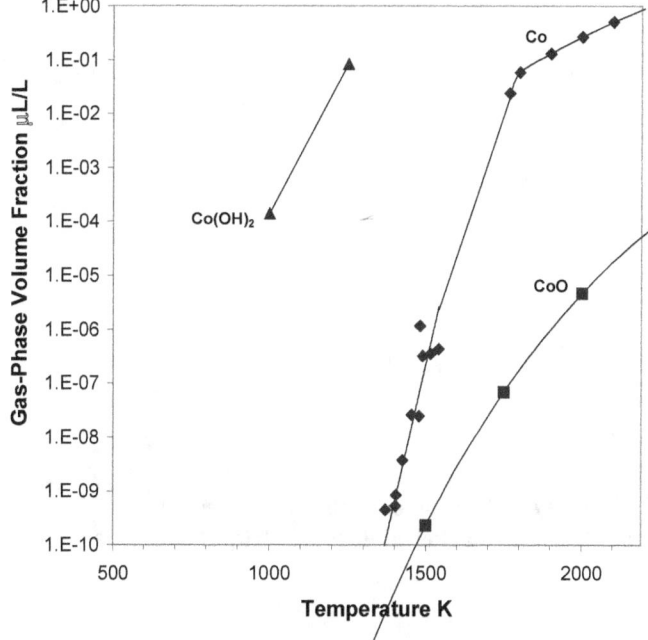

Figure 31 - Co species gas-phase volume fraction at equilibrium over the condensed-phase (at 1 atm), i.e., vapor pressure, Co, CoO, and Co(OH)$_2$, from (Nesmeianov, 1963; Gurvich et al., 1993; Stull, 1947).

Cu

As illustrated in Table 1, copper salts have been used in experiments on engine knock, ignition suppression, radical recombination above flat flames, as well as in flame screening tests. The compounds are added either as aqueous sprays or directly as particles. There has been no mention in the literature of observed particle formation after decomposition of the additive, or of any loss of effectiveness at higher mole fraction. Only one of these studies ran experiments over a range of additive mole fraction ((Rosser et al., 1963)), and in it, they attributed changes in the performance with additive mole fraction to heating effects of the particles (not condensation of active species or saturation of the radical recombination cycle). Hence, it is difficult to assess from previous work if condensation may be possible for copper-based inhibitors. To assess the condensation potential, we have assembled the available vapor pressure data for copper compounds which may exist in flames (see Figure 32). Although data are not available for other possible hydrides, oxides and hydroxides (e.g., CuH, CuOH) CuO and Cu_2O have quite high vapor pressures, and that of Cu is also relatively high at flame temperatures.

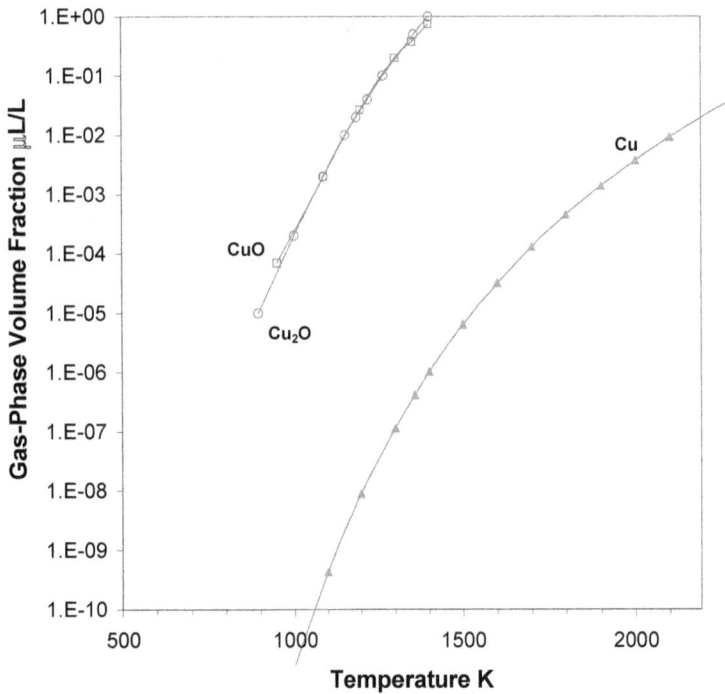

Figure 32 - Cu species gas-phase volume fraction at equilibrium over the condensed-phase (at 1 atm), i.e., vapor pressure, Cu, CuO, and Cu_2O, from {Barin, Knacke, et al. 1973 #710}{Barin, Knacke, et al. 1977 #720}.

Sb

Although the only flame data for antimony inhibition (Lask and Wagner, 1962) did not go to high enough concentrations to show loss of effectiveness, antimony may form condensed-phase particles in flames. In studies of polymers with added Sb_2O_3 and halogen, Fenimore and Marin (1966a) showed that the fire retardant effect increases linearly with Sb_2O_3 at low mass fractions of Sb_2O_3, but saturates at some value, above which further addition of Sb_2O_3 is ineffective. Hence, the fire retardancy effect of antimony shows a strong saturation, much like that for iron, manganese, and tin in cup-burner flames (Linteris, 2002). To illustrate this, the data in Figure 13 (CO_2 required for extinction with metal species added) is re-plotted below in Figure 33 in terms of the limiting oxygen index (based on N_2). The top of the figure shows the limiting oxygen index for polyethylene/halogen blends as a function of the volume fraction of the metallic inhibitor in the gas phase. The curves for Sb_2O_3 were calculated based on data available in Fenimore and Martin (1966b). The effectiveness of the antimony/halogen system saturates at Sb$_2$O$_3$ volume fractions near 400 µL/L (based on Sb). On the bottom of the figure, curves for Fe, Mn, and Sn added to the air stream of heptane-air cup burner flames are also shown (these data are obtained from (Linteris, 2002); although those experiments were conducted with CO$_2$ added as the diluent, the data were converted to an equivalent LOI with nitrogen diluent by correcting for the difference in heat capacity between N$_2$ and CO$_2$).

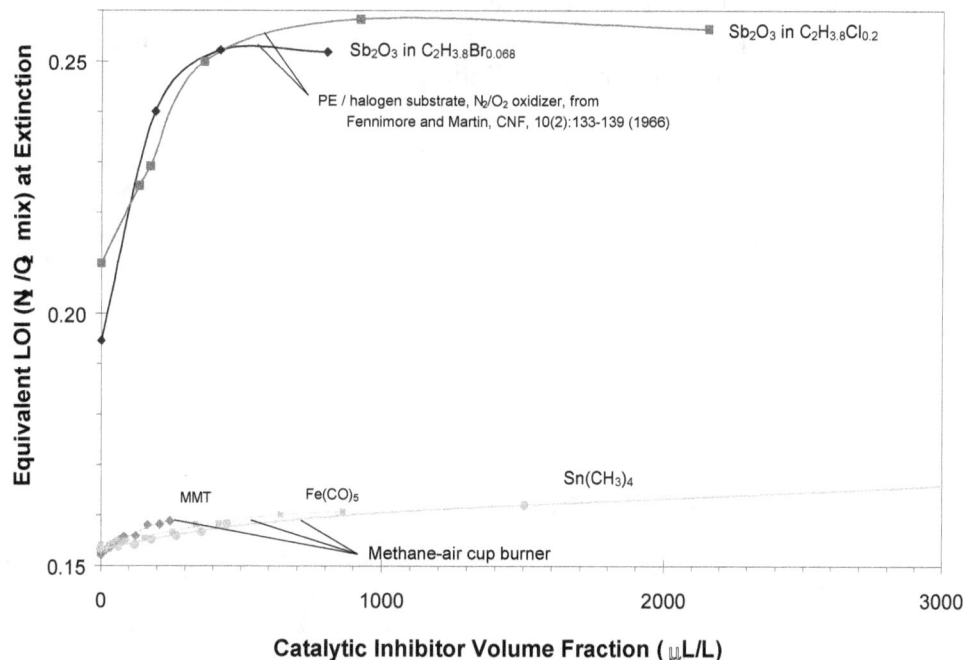

Figure 33 - Equivalent N$_2$/O$_2$ limiting oxygen index for extinction of polyethylene (PE) halogen blends or methane-air cup burner flames with MMT, Fe(CO)$_5$, and Sn(CH$_3$)$_4$.

Although the influence of Fe, Mn, and Sn on the LOI when added to the air stream of methane-air cup-burner flames is much weaker than that of Sb$_2$O$_3$ in halogenated polyethylene, a similar saturation behavior is observed. It would be of value to understand why saturation occurs in the antimony-halogen system, since it has not been explained in the literature.

Flame Inhibition and Loss-of-Effectiveness Summary

The current state of understanding of the inhibition of flames by metal compounds is summarized in Table 3. The inhibition potential is summarized in terms of the known inhibition behavior, the detection of metal-containing species in flames, and the state of development of a kinetic description of the inhibition. The knowledge of the potential for loss-of-effectiveness is characterized in terms of the demonstrated loss of effectiveness in flame systems, the presence of particles, and the availability of vapor pressure data for the condensed-phase metal-containing species. The *quality* of the information is rated as high, medium, low, and none. It should be noted however, that this is a relative scale. Even for elements such as iron, which is the most extensively studied, there are fragmented data on the gas-phase species in flames, the gas-phase kinetic mechanism has many estimated rate constants and the dominant inhibition cycles were assumed, and vapor pressure data for some of the important intermediates are not available. As the table illustrates, most of the information needed to accurately predict both the inhibition potential as well as the potential for loss-of-effectiveness is incomplete. (Note again that the rating in the Inhibition Potential column is the *quality* of the data about the inhibition potential, not the inhibition ability itself)

Table 3 – Current state of knowledge relevant to inhibition potential of metals, and potential loss of effectiveness due to condensation.   Key: ■ - high, ▨ - medium, ░ - low, -none.

Element	Inhibition Information			Condensation Information		
	Inhibition Potential	Gas-phase Species Identified	Kinetic Mechanism State of Development	Demonstrated Loss-of-Effectiveness	Experimental Evidence of Particles	Experimental Vapor Pressure Data
Cr	high	low	low	high	high	low
Pb	high	low	low	high	high	low
Fe	high	medium	low	high	high	low
Mn	high	medium	low	high	low	low
Ni	low					
W	low		low			low
Mo	low		low			low
Sn	low	high	low	high	low	low
Co	low	high	low			
Ti	low				low	
Ge	low					
Sb	low	low	low	low		
Te	low					
Tl	low					
Bi	low					
Cu	low	low				low
U	low					
Zn	low					
La	low					
Th	low					
Se	low					

Approaches for Ascertaining the Potential of Metal-containing Compounds

Actually calculating the condensation potential of metal inhibitors in flames is quite challenging. To do this in a *quantitative* way requires knowledge of the species present in the flame, the inhibition mechanism with its associated rate constants, the vapor pressures of the metal species, the condensation, agglomeration, and re-evaporation rates for the particles, and the actual species present in the particles. This is far from being achieved even for iron, the most highly studied metal.

An alternative way to assess the potential for condensation of metal inhibitors in flame systems may be simple screening experiments with cup-burners. In order to determine both the variation with concentration, as well as the potential benefit of using the metal compounds with inert agents, tests need to be conducted with the added metal compound and a secondary inert agent added together. In such a test, described above and in the literature (Linteris et al., 2002; Linteris and Rumminger, 2002) (Shmakov et al., 2004) the amount of a diluent (e.g., CO_2, or N_2) added to the air stream necessary to cause blow-off of the cup-burner flame is determined, both with and without the metal agent added at increasing concentration in the oxidizer stream. Since, to some extent, the flames resemble the low-strain conditions of fires, there is expected to be much higher correlation between the behavior in this system and in actual suppressed fires. By adding the metal compounds in the form of organometallic agents (rather than metal salts), the complicating effects of particle evaporation and decomposition are avoided.

A more empirical approach to evaluating metal-containing compounds would be to add them to a solid propellant-based extinguisher. There is inert gas in the effluent, as well as fragmented metal-containing species, so that by proper application to an appropriate fire, the utility of the metal additive could be assessed at larger scale. The challenge would be to find a configuration which has the proper sensitivity to the inert gas flow rate so that the effects of the additive could be quantified.

Proposed Approaches for Overcoming Condensation

There may exist approaches for overcoming the loss of effectiveness of metal compounds due to condensation of metal oxides. For example, halogens could be used to attack the metal oxide and provide metal-halogen species in the gas phase. As described by Hastie and co-workers (Hastie and McBee, 1975), the halogen in the antimony-halogen fire retardant system act to release the antimony from the condensed phase through a series of halogenation steps involving successive oxychloride phases (1973b). Bromine has been used in the past to etch off the lead oxide deposits on engine valves, and halogen is used to remove the oxide coating from the incandescent filament in quartz-halogen light bulbs. It is clear that halogens can release metals from solid oxides, and it may be possible to use this property to re-introduce the metals from the condensed oxide into the gas-phase where they can again inhibit the flame. Finally, the use of carboxylic acids as extenders of antiknock agents (Richardson et al., 1962) (through the formation of metal salts which can persist in the gas phase) leads naturally to the question of whether such an approach would work for metal-based fire suppressants as well.

Conclusions

An extensive review of the literature concerning flame inhibition by metals has been performed. The previous work has been considered with respect to the potential of the metals for flame inhibition, as well as the potential for loss of effectiveness due to condensation of the active species to particles. A list of metals with demonstrated flame inhibition potential has been compiled, along with the type of experiment used to recommend the metal, and the performance of the metal with respect to that of CF_3Br (neglecting any potential condensation of metal species) has been estimated. The mechanism of inhibition of iron compounds in flames has been described, as well as the mechanism of the loss of effectiveness of iron for some flame systems. The equivalent flame inhibiting species of the other metals has been suggested (when possible), and kinetic mechanisms have been outlined (when they have been proposed in the literature). The potential of the metals other than iron to lose effectiveness through condensation has been estimated, based on consideration of the performance of the agent as a function of its concentration, the observation of particles in the flame, or the vapor pressures of the active metal oxides and hydroxides (when the active species are known and the data are available).

This work has uncovered several additional metals which are likely to be effective at low concentrations, but which have not been suggested in previous reviews of metal flame inhibition. These are tungsten, molybdenum, and cobalt.

Most of the metals discussed have shown a possibility to condense to particles in some flame systems. These metals include Cr, Pb, Mn, Sn, and Sb. Were this condensation to occur in practical fires, compounds of these metals would likely not demonstrate fire suppression capabilities commensurate with their high flame velocity inhibition efficiency that was observed when added at low concentration to laboratory premixed flames. For other metals (W, Mo, Co, and Cu), there is no direct evidence of the potential for loss of effectiveness, and the vapor pressures of the suggested flame-quenching species (for which data are available) are reasonably high. For Co, the monoxide CoO has a lower vapor pressure and this is likely to be an important intermediate in the inhibition cycle. This can be taken as evidence for the potential for condensation. Nonetheless, the potential for condensation of metal species really depends upon the local super-saturation ratio in the inhibited flame. Calculation of the super-saturation ratio depends upon knowing both the detailed kinetic mechanism of inhibition as well as the vapor pressure of all of the intermediate species. Further, the kinetics of the condensation (and potential re-evaporation of particles) will be highly dependent on the flow-field of the particular flame system to be extinguished. Thus, *prediction* of the potential loss of effectiveness due to condensation is presently difficult.

To assess both the inhibition potential as well as the potential for loss of effectiveness for metal-based agents, two experimental approaches have been offered. Finally, to overcome the loss of effectiveness, the use of agents (halogens or carboxylic acids), which would keep a higher fraction in the metal in the gas phase, has been suggested.

Acknowledgements

Helpful conversations with Wing Tsang, Kermit Smyth, Fumi Takahashi, Dick Gann, George Mulholland, Vish Katta, and Bill Grosshandler are gratefully acknowledged. The author is indebted to Tania Ritchie, Marc Rumminger, and Valeri Babushok for their previous efforts in this research, upon which many of the current results depend. The assistance of Valeri Babushok in providing vapor pressure data for some of the metal species is gratefully noted. This research is part of the Department of Defense's Next Generation Fire Suppression Technology Program, funded by the DoD Strategic Environmental Research and Development Program under contract number W74RDV30663237.

References

Babushok, V. and Tsang, W. (2000). Inhibitor Rankings for Hydrocarbon Combustion. *Combustion and Flame,* **123,** 488.

Babushok, V., Tsang, W., Linteris, G. T., and Reinelt, D. (1998). Chemical Limits to Flame Inhibition. *Combustion and Flame,* **115,** 551.

Barin, Ihsan. and Knacke, O. (1973). *Thermochemical properties of inorganic substances* Berlin, New York, Springer-Verlag.

Barin, Ihsan., Knacke, O., and Kubaschewski, O. (1977). *Thermochemical properties of inorganic substances :* Springer-Verlag, Berlin ; New York.

Bonne, U., Jost, W., and Wagner, H. G. (1962). Iron Pentacarbonyl in Methane-Oxygen (or Air) Flames. *Fire Research Abstracts and Reviews,* **4,** 6.

Buelwicz, E. M., James, C. G., and Sugden, T. M. (1956). *Proceedings of the Royal Society A,* **235,** 89.

Bulewicz, E. M., Jones, G., and Padley, P. J. (1969). Temperature of metal oxide particles in flames. *Combustion and Flame,* **13,** 409.

Bulewicz, E. M. and Padley, P. J. (1971a). Catalytic Effect of Metal Additives on Free Radical Recombination Rates in $H_2+O_2+N_2$ Flames. *Proceedings of the Combustion Institute,* **13,** 73.

Bulewicz, E. M. and Padley, P. J. (1971b). Photometric investigations of the behavior of chromium additives in premixed H2+O2+N2 flames. *Proceedings of the Royal Society London A.,* **323,** 377.

Bulewicz, E. M., Padley, P. J., Cotton, D. H., and Jenkins, D. R. (1971). Metal-Additive-Catalysed Radical-Recombination Rates in Flames. *Chemical Physics Letters,* **9,** 467.

Bulewicz, E. M. and Sugden, T. M. (1956a). Determination of the dissociation constants and heats of formation of molecules by flame photometry; Part 3.-Heat of formation of cuprous hydroxide. *Transactions of the Faraday Society,* **52,** 14811488.

Bulewicz, E. M. and Sugden, T. M. (1956b). Determination of the dissociation constants and heats of formation of molecules by flame photometry; Part 2.-Heat of formation of cuprous hydride. *Transactions of the Faraday Society,* **52,** 1475.

Chamberlain, G. H. N. and Walsh, A. D. (1952). *Proceedings of the Royal Society A,* **215,** 175-.

Cheaney, D. E., Davies, D. A., Davis, A., Hoare, D. E., Protheroe, J., and Walsh, A. D. (1959). Effects of Surfaces on Combustion of Methane and Mode of Action of Anti-Knock Containing Metals, *Seventh Symposium (International) on Combustion,* The Combustion Institute, p. 183.

Cotton, D. H., Friswell, N. J., and Jenkins (1971). The suppression of soot emission from flames by metal additives. *Combustion and Flame,* **17,** 87.

Cotton, D. H. and Jenkins, D. R. (1971). Catalysis of Radical-Recombination Reactions in Flames by Alkaline Earth Metals. *Transactions of the Faraday Society,* **67,** 730.

deWitte, M., Vrebosch, J., and van Tiggelen, A. (1964). Inhibition and extinction of premixed flames by dust particles. *Combustion and Flame,* **9,** 257.

Dolan, J. E. and Dempster, P. B. (1955). The suppression of methane-air ignitions by fine powders. *Journal of Applied Chemistry,* **5,** 510.

Erhard, K. H. L. and Norrish, R. G. W. (1956). Studies of knock and antiknock by kinetic spectroscopy. *Proc. Roy. Soc. (London) A,* **234,** 178.

Fallis, S., Reed, R., Lu, Y.-C., Wierenga, P. H., and Holland, G. F (2000). Advanced Propellant/Additive Development for Fire Suppressing Gas Generators, *Halon Options Technical Working Conference,* 361.

Fallon, G. S., Chelliah, H. K, and Linteris, G. T. (1996). Chemical Effects of CF_3H in Extinguishing Counterflow $CO/Air/H_2$ Diffusion Flames, *Proceedings of the Combustion Institute, Vol. 26,* The Combustion Institute, p. 1395.

Fenimore, C. P. and Jones, G. W. (1966). Modes of Inhibiting Polymer Flammability. *Combustion and Flame,* **10,** 295.

Fenimore, C. P. and Martin, F. J. (1966a). Flammability of Polymers. *Combustion and Flame,* **10,** 135.

Fenimore, C. P. and Martin, F. J. (1966b). Candle-type test for flammability of polymers. *Modern Plastics,* 141.

Fenimore, C. P. and Martin, F. J. (1972), in *The mechamisms of pyrolysis, oxidation and burning of organic materials* (Wall, L. A., Ed.), National Bureau of Standards, Washington, D.C., pp. 159-170.

Gann, R. G. (2004). *FY2003 Annual Report -- Next Generation Fire Suppression Technology Program (NGP),* National Institute of Standards and Technology, NIST Technical Note 1457.

Grmiley, R. T., Burns, R. P., and Inghram, M. G. (1961). Thermodynamics of the vaporization of Cr_2O_3: dissociation energies of CrO, CrO_2, and CrO_3. *Journal of Chemical Physics,* **34,** 664.

Gurvich, L. V., Iorish, V. S., Chekhovskoi, D. V., Ivanisov, A. D., Proskurnev, A. Yu., Yungman, V. S., Medvedev, V. A., Veits, I. V., and Bergman, G. A. (1993). *IVTHANTHERMO - Database on Thermodynamic Properties of Individual Substances,* Institute of High Temperatures, Moscow.

Hastie, J. W. (1973a). Mass spectroscopic studies of flame inhibition: analysis of antimony trihalides in flames. *Combustion and Flame,* **21,** 49.

Hastie, J. W. (1973b). Molecular Basis of Flame Inhibition. *Journal of Research of the National Bureau of Standards - A. Physics and Chemistry,* **77A,** 733.

Hastie, J. W. and McBee, C. L. (1975), in *Halogenated Fire Suppressants* (Gann, R. G., Ed.), American Chemical Society, Washington, DC, pp. 118-148.

Howard, J. B. and Kausch, W. J. (1980). Soot Control by Fuel Additives. *Progress in Energy and Combustion Science*, **6,** 263.

James, C. G. and Sugden, T. M. (1956). *Proceedings of the Royal Society A*, **248,** 238.

Jensen, D. E. and Jones, G. A. (1974). Catalysis of Radical Recombination in Flames by Iron. *Journal of Chemical Physics*, **60,** 3421.

Jensen, D. E. and Jones, G. A. (1975). Mass-Spectrometric Tracer and Photometric Studies of Catalyzed Radical Recombination in Flames. *Journal of the Chemical Society-Faraday Transactions I*, **71,** 149.

Jensen, D. E. and Jones, G. A. (1976). Aspects of Flame Chemistry of Cobalt. *Journal of the Chemical Society-Faraday Transactions I*, **72,** 2618.

Jensen, D. E. and Webb, B. C. (1976). Afterburning Predictions for Metal-Modified Propellant Motor Exhausts. *AIAA Journal*, **14,** 947.

Jost, W., Bonne, U., and Wagner, H. G. (1961). *Chemical & Engineering News*, **39,** 76.

Katta, V. R., Takahashi, F., and Linteris, G. T. (2003a). Extinction Characteristics of Cup-Burner Flame in Microgravity, *41st Aerospace Sciences Meeting and Exhibit*, AIAA, p. AIAA Paper No. 2003.

Katta, V. R., Takahashi, F., and Linteris, G. T. (2003b). Numerical Investigations of CO2 as Fire Suppressing Agent, *Fire Safety Science: Proc. of the Seventh Int. Symp.*, Int. Assoc. for Fire Safety Science, p. 531.

Kellogg, C. B. and Irikura, K. K. (1999). Gas-Phase Thermochemistry of Iron Oxides and Hydroxides:Portrait of a Super-Efficient Flame Suppressant. *Journal of Physical Chemistry a*, **103,** 1150.

Kennedy, I. M., Zhang, Y., Jones, A. D., Chang, D. P. Y., Kelly, P. B., and Yoon, Y. (1999). Morphology of Chromium Emissions From a Laminar Hydrogen Diffusion Flame. *Combustion and Flame*, **116,** 233.

Kovarik, B. (1994). Charles F. Kettering and the Development of Tetraethyl Lead in the Context of Alternative Fuel Technologies, *Fuels and Lubricants*, Society of Automotive Engineers, p.

Kuppu Rao, V. and Prasad, C. R. (1972). Knock suppression in petrol engines. *Combustion and Flame*, **18,** 167.

Lask, G. and Wagner, H. G. (1962). Influence of Additives on the Velocity of Laminar Flames. *Proceedings of the Combustion Institute*, **8,** 432.

Linteris, G. T. (2002). Extinction of Cup-Burner Diffusion Flames by Catalytic and Inert Inhibitors, *NRIFD*,

Linteris, G. T. and Chelliah, H. K. (2001). *Powder-Matrix Systems for Safer Handling and*

Storage of Suppression Agents, National Institute of Standards and Technology, NISTIR 6766.

Linteris, G. T., Katta, V. R., and Takahashi, F. (2004). Experimental and Numerical Evaluation of Metallic Compounds for Suppressing Cup-Burner Flames. *Combustion and Flame,* **138,** 78.

Linteris, G. T., Knyazev, K., and Babushok, V. (2002). Inhibition of Premixed Methane Flames by Manganese and Tin Compounds. *Combustion and Flame,* **129,** 221.

Linteris, G. T. and Rumminger, M. D. (2002). Particle Formation in Laminar Flames Inhibited by Metals, *Western States Section Meeting, The Combustion Institute* Paper 030.

Linteris, G. T., Rumminger, M. D., Babushok, V., Chelliah, H., Lazzarini, A. K., and Wanigarathne, P. (2002). *Effective Non-Toxic Metallic Fire Suppressants,* National Institute of Standards and Technology, NISTIR 6875.

Linteris, G. T., Rumminger, M. D., Babushok, V., and Tsang, W. (2000a). Flame Inhibition by Ferrocene and Blends of Inert and Catalytic Agents. *Proceedings of the Combustion Institute,* **28,** 2965.

Linteris, G. T., Rumminger, M. D., and Babushok, V. I. (2000b). Premixed Carbon Monoxide-Nitrous Oxide-Hydrogen Flames:Measured and Calculated Burning Velocities With and Without $Fe(CO)_5$. *Combustion and Flame,* **122,** 58.

Lippincott, E. S. and Tobin, M. C. J. (1953). *Journal of the American Chemical Society,* **75,** 4141.

Lott, J. L., Christian, S. D., Sliepcevich, C. M., and Tucker, E. E. (1996). Synergism Between Chemical and Physical Fire-Suppressant Agents. *Fire Technology,* **32,** 260.

Macdonald, M. A., Gouldin, F. C., and Fisher, E. M. (2001). Temperature Dependence of Phosphorus-Based Flame Inhibition. *Combustion and Flame,* **124,** 668.

MacDonald, M. A., Jayaweera, T. M., Fisher, E. M., and Gouldin, F. C. (1998). Variation of Chemically Active and Inert Flame-Suppression Effectiveness with Stoichiometric Mixture Fraction, *Proceedings of the Combustion Institute, Vol. 27,* The Combustion Institute, p.

Martin, F. J. and Price, K. R. (1968). *Journal of Applied Polymer Science,* **12,** 143.

Matsuda, S. (1972). Gas Phase Homogeneous Catalysis in Shock Waves. II. The Oxidation of Carbon Monoxide by Oxygen in the Presence of Iron Pentacarbonyl. *Journal of Physical Chemistry,* **57,** 807.

Matsuda, S. and Gutman, D. (1971). Shock-tube study of C_2H_2-O_2 reaction. Acceleration of reaction in the presence of trace amounts of $Cr(CO)_6$. *Journal of Physical Chemistry,* **75,** 2402.

Miller, D. R., Evers, R. L., and Skinner, G. B. (1963). Effects of Various Inhibitors on Hydrogen-Air Flame Speeds. *Combustion and Flame,* **7,** 137.

Miller, W. J. (1969). Inhibition of Low Pressure Flames. *Combustion and Flame,* **13,** 210.

Morrison, M. E. and Scheller, K. (1972). The Effect of Burning Velocity Inhibitors on the Ignition of Hydrocarbon-Oxygen-Nitrogen Mixtures. *Combustion and Flame,* **18,** 3.

Muraour, H. (1925). *Chem. et Industr.,* **14,** 1911.

Noto, T., Babushok, V., Hamins, A., and Tsang, W. (1998). Inhibition effectiveness of halogenated compounds. *Combustion and Flame,* **112,** 147.

Padley, P. J. and Sugden, T. M. (1959). Determination of the dissociation constants and heats of formation of molecules by flame photometry; Part 6.-stability of MnO and MnOH and their mechanisms of formation. *Transaction of the Faraday Society,* **55,** 2054.

Park, K., Bae, G. T., and Shin, K. S. (2002). The addition effect of Fe(CO)5 on methane ignition. *Bulletin of the Korean Chemical Society,* **23,** 175.

Reinelt, D. and Linteris, G. T. (1996). Experimental Study of the Inhibition of Premixed and Diffusion Flames by Iron Pentacarbonyl. *Proceedings of the Combustion Institute,* **26,** 1421.

Richardson, W. L., Ryason, P. R., Kautsky, G. J., and Barusch, M. R. (1962). Organolead antiknock agents--their performance and mode of action. *Proceedings of the Combustion Institute,* **9,** 1023.

Rosser, W. A, Inami, S. H., and Wise, H. (1959). *Study of the Mechanisms of Fire Extinguishment of Liquid Rocket Propellants,* WADC Technical Report 59-206.

Rosser, W. A, Inami, S. H., and Wise, H. (1963). The Effect of Metal Salts on Premixed Hydrocarbon-Air Flames. *Combustion and Flame,* **7,** 107.

Rumminger, M. D. and Linteris, G. T. (2000a). Numerical Modeling of Counterflow Diffusion Flames Inhibited by Iron Pentacarbonyl, *Fire Safety Science: Proc. of the Sixth Int. Symp.,* Int. Assoc. for Fire Safety Science, p. 289.

Rumminger, M. D. and Linteris, G. T. (2000b). Inhibition of Premixed Carbon Monoxide-Hydrogen-Oxygen-Nitrogen Flames by Iron Pentacarbonyl. *Combustion and Flame,* **120,** 451.

Rumminger, M. D. and Linteris, G. T. (2000c). The Role of Particles in the Inhibition of Premixed Flames by Iron Pentacarbonyl. *Combustion and Flame,* **123,** 82.

Rumminger, M. D. and Linteris, G. T. (2002). The Role of Particles in the Inhibition of Counterflow Diffusion Flames by Iron Pentacarbonyl. *Combustion and Flame,* **128,** 145.

Rumminger, M. D., Reinelt, D., Babushok, V., and Linteris, G. T. (1999). Numerical Study of the Inhibition of Premixed and Diffusion Flames by Iron Pentacarbonyl. *Combustion and Flame,* **116,** 207.

Shmakov, A. G., Korobeinichev, O. P., Shvartsberg, V. M., Knyazkov, D. A., Bolshova, T. A., and Rybitskaya, I. V. (2004). Inhibition of Premixed and Non-Premixed Flames with Phosphorus-Containing Compounds. *Proceedings of the Combustion Institute,* **30,** in press.

Stull, D. R. (1947). Vapor Pressure of Pure Substances Organic Compounds. *Ind. Eng. Chem.,* **39,** 517.

Suyuan Y., Jones, A. D., Chang, D. P. Y., Kelly, P. B., and Kennedy, I. M. (1998). The Transformation of Chromium in a Laminar Premixed Hydrogen-Air Flame. *Proceedings of the*

Combustion Institute, **27,** 1639.

Tapscott, R. E., Sheinson, R. S., Babushok, V., Nyden, M. R., and Gann, R. G (2001). *Alterntive Fire Suppressant Chemicals: A Research Review with Recommendations,* NIST TN 1443.

Tischer, R. L. and Scheller, K. (1970). The behavior of uranium oxide particles in reducing flames. *Combustion and Flame,* **15,** 199.

Vanpee, M. and Shirodkar, P. (1979). A Study of Flame Inhibition by Metal Compounds. *Proceedings of the Combustion Institute,* **17,** 787.

Walsh, A. D. (1954), in *Six Lectures on the Basic Combustion Process* (Ethyl Corp., New York, NY, pp.

Westblom, U., Fernandezalonso, F., Mahon, C. R., Smith, G. P., Jeffries, J. B., and Crosley, D. R. (1994). Laser-Induced Fluorescence Diagnostics of a Propane Air Flamewith a Manganese Fuel Additive. *Combustion and Flame,* **99,** 261.

Zimpel. C.F. and Graiff, L. B. (1967). An Electron Microscopic Study of Tetraethyllead Decomposition in an Internal Combustion Engine, *Eleventh Symposium (International) on Combustion,* The Combustion Institute, p. 1015.